Roger Moser

Strategic Purchasing and Supply Management

GABLER EDITION WISSENSCHAFT

Einkauf, Logistik und Supply Chain Management

Herausgegeben von
Professor Dr. Christopher Jahns

Die Schriftenreihe stellt den State-of-the-art betriebswirtschaftlicher Forschung am Supply Management Institute SMI™ im Bereich Einkauf, Logistik und Supply Chain Management dar. Die Verbindung von Theorie und Praxis steht dabei ebenso im Vordergrund wie die internationale Ausrichtung und die unmittelbare Verknüpfung der Themen Einkauf, Logistik und Supply Chain Management.

Roger Moser

Strategic Purchasing and Supply Management

A Strategy-Based Selection of Suppliers

With a foreword by Prof. Dr. Christopher Jahns

Deutscher Universitäts-Verlag

Bibliografische Information Der Deutschen Nationalbibliothek
Die Deutsche Nationalbibliothek verzeichnet diese Publikation in der Deutschen Nationalbibliografie; detaillierte bibliografische Daten sind im Internet über <http://dnb.d-nb.de> abrufbar.

Dissertation European Business School Oestrich-Winkel, 2006

D 1540

1. Auflage August 2007

Lektorat: Frauke Schindler / Nicole Schweitzer

Der Deutsche Universitäts-Verlag ist ein Unternehmen von Springer Science+Business Media.
www.duv.de

Umschlaggestaltung: Regine Zimmer, Dipl.-Designerin, Frankfurt/Main
Gedruckt auf säurefreiem und chlorfrei gebleichtem Papier
Printed in Germany

ISBN 978-3-8350-0658-4

Foreword of the Editor

Strategic management has neglected purchasing and supply management (PSM) for a long time - not only from an academic perspective but also from a practitioner's point of view. One reason for this might be due to the fact that only a few universities in Europe and the United States have significant faculty resources for teaching and conducting research in PSM. Therefore, most students do not have contact to topics such as supplier management, supply market analysis or specification management during their undergraduate or graduate studies.

As a result, PSM is still not considered a strategic function in many companies, although on average more than half of a company's value proposition comes from its suppliers. In today's economy, companies which neglect the strategic management of their supply base fall behind their competitors and in many boardrooms PSM is currently on top of the agenda. PSM can only create the competitive advantages necessary to stay ahead of the competition when a company's supply base is managed according to its overall strategic objectives.

The analysis of the relationships between business strategy priorities and PSM strategy is the focus of Dr. Roger Moser's dissertation. His studies analyze in detail how business strategies influence PSM strategy, in particular supplier selection, and how a strategy-based selection of suppliers can generate competitive advantages in PSM and for the company. His work contributes to a research stream which aims at integrating PSM more closely into strategic management.

Based on Dr. Moser's work, companies can understand how their business strategy priorities are linked to their PSM strategies and supplier selection criteria. He further develops a PSM strategy concept which enables companies to break down strategic priorities from a business strategy level to a PSM level and to define appropriate actions when dealing with suppliers, supply markets and internal customers. Based on empirical data, Dr. Moser also shows which supplier selection criteria companies should prioritize in order to reduce costs, improve sales or reduce risks.

In summary, Dr. Moser provides a comprehensive, methodical and highly interesting dissertation. In addition to his ambitious research concerning new innovations in PSM, his interest in transferring scientific methods in PSM is inspiring. I was quite impressed not only how quickly yet thoroughly the dissertation project was completed, but also by Dr. Moser's ability to work under pressure. While conducting research for this project, Dr. Moser worked as Project Manager for a worldwide PSM study conducted together with McKinsey & Co.

I wish Dr. Moser all the best for his professional future. His efforts were recently rewarded with a Junior Professorship appointment for Global Supply Networks only a few months after finishing his dissertation. This appointment is unique since Dr. Moser actively participated for 12 months in the academic establishment of the BMW-SMI Endowed Chair for Purchasing and Supply Management in China and will now take over responsibilities for 2 years in the academic establishment of an Endowed Chair for Sourcing and Supply Management at the renowned Indian Institute of Management Bangalore. The upcoming stay in Bangalore will enable Dr. Moser to further develop the research field of PSM – his passion – with highly talented Indian academics.

Univ.-Prof. Dr. Christopher Jahns

Foreword of the Author

The writing process of a dissertation is often associated with many hours of reflection, isolation from the real world and diligence. The latter is certainly true for any dissertation written at the Supply Management Institute SMI of the European Business School (EBS). However, real isolation never really existed during my time as a doctoral candidate, neither personally nor intellectually. Many of my colleagues often stayed with me long after midnight and challenged my ideas and concepts – their moral support was greatly appreciated but in no respect expected. As in today's business world, where companies rarely succeed without strong partners, my dissertation would never have been realized without the support of many people. First of all, I owe many thanks to my supervisors, Univ.-Prof. Dr. Christopher Jahns and Univ.-Prof. Dr. Ronald Gleich, for their scientific guidance and professional support during my time as a doctoral candidate at the European Business School (EBS), International University Schloss Reichartshausen. They supported my proposal to investigate the relationships between business strategies and supplier selection practices in more detail because their academic and practice experience had shown that neither researchers nor practitioners were fully aware of the significance of this relationship. Therefore, the results of this doctoral thesis will hopefully contribute to the better integration of purchasing and supply management (PSM) into the strategic management of companies.

Neither my university studies nor my dissertation would have been possible without the support of my parents, Guido and Beatrice Moser-Hauser. They have always supported me in whatever I wanted to do, as long as I worked hard. The example they set was essential to master the challenge of writing this dissertation. Moreover, I want to thank my wife, Nadja Moser-Stübi, for all her support during the last seven years. Her support especially during my time as a doctoral candidate was essential for the realization of my objectives.

My dissertation would also not have been possible without the support and critical comments of the professors: Prof. Dr. Evi Hartmann, Prof. Dr. Inga-Lena Darkow, Prof. Dr. Stefan Walter and Univ.-Prof. Dr. Michael Henke at the Supply Management Institute SMI at the European Business School (EBS), and my colleagues Dr. Martin Lockström, Gernot Kaiser, Julia Wolf, Aiko Entchelmeier, Andreas Potzner and many more. Finally, I want to thank Konrad Bänziger and his colleagues of the Supply Management Group SMG for their support and critical comments during the dissertation writing process.

Dr. Roger Moser

Content

List of figures

List of tables

List of abbreviations

AHP	Analytical Hierarchy Process
AMOS	Analysis of Moment Structures
ANP	Analytical Network Process
ASEAN	Association of Southeast Asian Nations
AVE	Average Variance Extracted
CR	Composite Reliability
e.g.	exempla gratia (for example)
EU	European Union
H&M	Hennes & Mauritz
i.e.	id est (this means)
IMP	Industrial Marketing and Purchasing
JIT	Just-in-Time
LISREL	Linear Structural Relationships
NAFTA	North American Free Trade Agreement
PLS	Partial Least Squares
PSM	Purchasing and Supply Management
R&D	Research & Development
SLV	Supplier Lifetime Value
TQM	Total Quality Management
TSLV	Target Supplier Lifetime Value
TVE	Total Variance Explained
VIF	Variance Inflation Factor
X-BSC	Cross-Balanced Scorecard

1 Problem situation and research approach

In the 1990s, the internal business processes of companies had been reengineered. Today, researchers and practitioners are predicting that the coming decades will be the era of reengineering the entire value chain – from the initiation of the customer request through design, manufacturing, logistics, and service to the final customer again. The winners in the future marketplace will be those linked companies which can combine their internal advantages with those of their suppliers to a powerful value chain that is faster, more agile and innovative, and ultimately more profitable than other competing value chains (Burt *et al.*, 2002; Spekman *et al.*, 1999).

With this erosion of competitive advantages of single companies over recent years, many capabilities of companies alone do no longer serve as a competitive weapon in the world market. Therefore, companies must find a new set of competitive advantages and purchasing and supply management (PSM[1]), along with its supply base, holds great potential for reaching the necessary improvements (Drucker, 1982; Hogan *et al.*, 2001; Jap, 1999). Top-managers should recognize the critical role PSM can play in shaping the competitiveness of their companies together with their suppliers in the future in terms of both bottom and top line results (Monczka *et al.*, 2002; Watts *et al.*, 1995).

An illustrative example for the generation of competitive advantages with suppliers is Toyota which has made more profits in 2003 than its three largest competitors put together. In contrast to many other automotive companies, Toyota does not keep its suppliers at arm's length guarding its internal knowledge. On the contrary, it embraces its suppliers and encourages knowledge sharing with them by establishing networks that facilitate the exchange of information. This provides Toyota with a significant competitive advantage over other automotive companies not integrating their suppliers (Dyer *et al.*, 2004). Another example of a successful buyer-supplier relationship is the computer systems company Dell Inc. where suppliers are the very lifeblood of the company's business. At Dell Inc. effective knowledge sharing with its supply partners is crucial for the company's success (Agrawal *et al.*, 2001; Magretta, 1998). The experiences of Toyota or Dell strongly suggest that competitive advantages can be achieved and sustained through PSM and its supplier management.

1 In this doctoral thesis, purchasing and supply management (PSM) is used as a holistic term including purchasing, procurement, sourcing and supply management. If other terms than PSM are used in the text, they stem from the analyzed original literature closely related to the statements made. This should support the readers' understanding of the still fragmented and different use of the most relevant terms (for a similar use compare e.g. Baily, 2001; Carter *et al.*, 1998; Dobler *et al.*, 1996; Ellram *et al.*, 2002b; Leenders *et al.*, 2002).

In the following two subchapters, the role of suppliers in generating competitive advantages is analyzed and the research approach of this doctoral thesis is presented.

1.1 The role of strategic suppliers in generating competitive advantages

Recent developments in the world's economy foster a significant change in the strategic role of suppliers for companies. Firstly, competition has intensified on a global scale over the last twenty years. Important factors are deregulation, the emergence of new free trade zones such as ASEAN (Association of South-East Asian Nations), NAFTA (North American Free Trade Agreement) or EU (European Union), improved transportation possibilities, intercultural homogenization, and sophisticated information technology (van Weele *et al.*, 1996; Vonderembse, 2002). Exploiting the new information and logistics linkages, companies are relocating and reconfiguring their activities into global value chains based on least cost and greatest expertise. For instance, some companies have become global in the way their R&D is networked in order to apply simultaneous engineering. As a result, design concepts are detailed by low-cost workers such as Indian engineers and subsequently switched over to high-productivity locations such as Europe, the U.S. or Singapore for final tests. The necessary components are again produced and assembled in low-cost countries such as China, Taiwan or Eastern Europe. Finally, most products are sold in Europe, in North America or in the emerging markets (van Weele *et al.*, 1996). These developments imply a drastic increase in the complexity and global scale of buyer-supplier relationships.

As another outcome of these developments, outsourcing (Kotabe, 2002) and offshoring (Farrell, 2005) concepts have recently gained more attention of senior and top management. These concepts imply that based on internal and external cost/price analyses, carried out in the context of competitive benchmarking programs, companies discover that particular activities can no longer be carried out competitively within their structures. Internal production costs appear to be much higher than the costs of external suppliers. This prompts companies to start make-or-buy analysis programs. As a result, companies focus their manufacturing processes on higher assembly levels, or modular components are sourced from a limited number of preferred suppliers while the focal company focuses on design, marketing and sales (van Weele, 2002). The trends of outsourcing and offshoring increase the reliance and dependence on suppliers and also the importance of an effective and efficient supplier management (Gottfredson *et al.*, 2005; Kannan *et al.*, 2002; Norrman *et al.*, 2004).

Another management concept which has emerged during recent years is mass customization (Kemppainen *et al.*, 2003; Normann *et al.*, 1993; Piller, 2001; Pine, 1993; Riemer *et al.*, 2002; Womack *et al.*, 1990). As customers require more customized products and services, PSM is increasingly required to select and develop its supply base according to their needs (Guinipero *et al.*, 2004b). The development of mass customization concepts has been triggered by more demanding customers and continuously changing preferences. In the past, customers had judged the value of a product or service on the basis of some combination of quality and price. The costumers of tomorrow will employ an expanded concept of value that also encompasses convenience of purchase, after-sales service, uniqueness or reliability. The costumers will take charge of the business-to-consumer relationship: they will tell manufacturers what they want, when they want it, how they want it and what they are willing to pay (Claus *et al.*, 2003; van Weele *et al.*, 1996). Only with the support of their suppliers, companies will be able to customize products or services for individual customers in high volumes and at relatively low costs (Leek, 2003; Monczka *et al.*, 2000b).

The pace of technology and product development has also dramatically increased the strategic role of PSM and its supply base (Cavinato *et al.*, 2000). The effective transfer of technology in a timely manner has become a critical factor of success. Concurrently, the costs and complexity of new technology development have continued to increase, and technology venturing and partnering with suppliers have emerged as a major possibility to maximizing commercial opportunities and technologies' value. The integration of external technology sources results in an acceleration of technology commercialization, improved cost-efficiency, stronger technology competencies and a greater scope of technology reach and customer intelligence (Monczka *et al.*, 2000a). In addition, the traditional sequential design-and-manufacture process is superseded by a parallel activity known as concurrent or simultaneous engineering (Hartmann *et al.*, 2005). For example, some engineering organizations are increasingly beginning to focus their efforts solely upon their own core technologies and specialties, and delegate the design and development of non-core technology parts to suppliers. The role of the suppliers is therefore evolving from one, solely providing components to a role that includes the provision of design information and development knowledge. Nowadays, engineers and designers often have to rely heavily upon suppliers for information and expertise throughout their own engineering and design processes (Culley *et al.*, 1999; Vonderembse, 2002).

To put it in a nutshell, the markets in which firms compete are increasingly influenced by international competition, demanding customers, rapid technological change, shorter product life cycles, advanced information technologies and new logistics solutions (Lowson, 2003). Consequently, companies and managers experience more and more pressure on margins and performance. These developments drive factors such as focusing on core competences (Prahalad *et al.*, 1990), outsourcing of strategically relevant activities or introducing mass customization concepts (Fredriksson *et al.*, 2003; Kannan *et al.*, 2002; Wagner, 2003). Consequently, companies turn to suppliers to get their support in implementing these concepts and achieving a stronger competitive position (Ganesan, 1994; Testore, 1998).

This increased interaction between buying firms and their suppliers results in less vertically integrated companies (Stuart, 1997). According to recent surveys, companies' degree of value added for products or services is constantly decreasing to a cross-industry level of less than 30% in 2007 (e.g. Jahns, 2005b; Sattler *et al.*, 2005). However, a lower level of vertical integration implies increased dependence on suppliers for timely delivery of quality products and services at competitive prices but also for process improvements and product innovations (Kannan *et al.*, 2002; Lemke *et al.*, 2000). This new existing interdependency between suppliers and buying firms is neither intrinsically good nor bad but it implies that a buying firm's ability to compete in its respective customer markets is impeded if the company experiences deficiencies in the integration of its suppliers' capabilities (Krause *et al.*, 1999).

Furthermore, the management of extended enterprises [2] will imply greater reliance on information technology, compatibility of planning and information systems, and a greater need for strategic sourcing. As a result, PSM professionals will require greater general management competences. PSM staff will pursue their work increasingly based on companies' strategic priorities and will be expected to respond pro-actively to market and strategy changes (Carter *et al.*, 1998). In the end, PSM performance measurement will be linked more closely to company objectives such as business growth, profitability, market share, and customer satisfaction.

These changes create a dynamic and vital PSM environment where companies need to deliver more value in new ways, to be faster to market, to become more flexible in responding to demand changes and to reduce costs in a sustainable way (Carter *et al.*, 2003). In order to achieve these higher performance levels, many companies have turned to their supply base in order to get access to capabilities they themselves can no longer provide or develop. This increases the need for higher levels of coordination among suppliers and buying firms and

[2] The concept of the extended enterprise was first developed by Chrysler Inc. which also granted a trademark for the term 'extended enterprise'. Chrysler Inc. defines 'extended enterprise' as extending business relationships by providing process management consultation and workshops to suppliers and supplier tiers in order to reduce cycle time, to minimize system cost, and to improve the quality of the goods or services provided by the suppliers (Dyer, 2000; Ericksen, 2001).

often leads to a different kind of relationship away from transactional approaches to more interaction between companies and their suppliers. By integrating the capabilities of other companies into their specific supply chain, buying firms can create unique value more efficiently and effectively (Katsikeas *et al.*, 2004).

This value is maximized when the supply chain acts in unison, almost as if it was one competitive unit in the marketplace (Rice *et al.*, 2001) but integrates more flexibility than single companies. Accordingly, the role of suppliers is evolving from one of solely providing components or products to a role that includes the provision of design, process management or innovation capabilities (Watts *et al.*, 1995). Today it is apparent that buyer-supplier interfaces often play a key role in the design and development of new products and that the mechanisms by which suppliers are successfully incorporated into, e.g., the engineering design process need to be better understood (Culley *et al.*, 1999). Based on this analysis of the current situation in the global competitive environment, buying firms are supposed to develop strategic approaches for their supplier management. Although there is a growing recognition that collaborative and adversarial relationships in business-to-business markets offer significant opportunities for companies to create competitive advantages and achieve superior results, integrated approaches for an overall strategy-based supplier management have not been developed yet and empirical studies dedicated to this problem are difficult to identify (Hewitt *et al.*, 2002; Jap, 1999; Walter *et al.*, 2001).

As a consequence, the problems in strategic supplier management faced by companies are numerous. First of all, suppliers and buyers need to understand how they can create and deliver value in business-to-business relationships. As for today, a majority of executives perceives suppliers from a cost, time and quality perspective, missing the fact that often 50-70% of a product or service value come from suppliers providing capabilities of pivotal importance for their business system. Only a few, if any company, can afford to overlook the additional value that suppliers' capabilities can contribute to cost reductions, sales increase or risk reduction (Carter *et al.*, 1998; Doyle, 1990).

The determination of this value creation in buyer-supplier relationships is still in its infancy and a sound understanding of the requirements from companies is a prerequisite for the development of reliable and valid strategic supplier management tools (Ulaga, 2003). Slowly but steadily, managerial emphasis in strategic supplier management is shifting from the evaluation of suppliers' current offers such as products and services to the evaluation of suppliers' capabilities and the value they add to buying firms' business (Anderson *et al.*, 1999;

Odgen *et al.*, 2005). These developments implicate dramatic changes pertaining to the selection and subsequent integration of suppliers. Supplier selection processes become increasingly important in order to exploit suppliers to their full potential (Culley *et al.*, 1999).

The challenge is to develop differentiated supplier strategies to support the buying firm's overall competitive and supply chain strategies (van Weele, 2002). Companies have to select those suppliers who support them best with their capabilities in implementing corporate strategies. This requires the build-up of optimally adapted supplier relationships on the continuum from collaborative partnerships to adversarial relationships (Hutchins, 2002; Wagner, 2001). In order to derive such benefits, companies must have a formal supplier strategy which is undertaken as part of the company's overall competitive strategy. They must put in place a systematic process to ensure that they are making the best competitive use of their internal core competencies and that they have access to the capabilities and technologies likely to properly complement their internal capabilities over the long term (Monczka *et al.*, 2000a).

One of the main problems with strategic supplier management is the missing linkage between corporate strategy, PSM strategy and supplier management. Many companies don't have their supplier strategy designed to effectively support the business mission. A sound understanding of the company's business mission and strategic priorities is vital for PSM in order to achieve competitive advantages with the support of its suppliers (Michiels, 2000; Moharty *et al.*, 1993). As a second challenge, suppliers' capabilities need to be assessed in a way that takes all relevant supplier performance potentials into account and links companies' strategic priorities and requirements more directly with suppliers' capabilities in order to choose the most suitable supplier or to develop the supplier with the best potential (Monczka *et al.*, 1994). For instance, while some companies may need suppliers with strong product innovation capabilities, others have to focus more on suppliers whose capabilities support their cost rationalization efforts (Araujo *et al.*, 1999). These different requirements towards suppliers also imply that no common definition of strategic suppliers can exist. Each company has to define its strategic suppliers based on their strategic priorities and their internal capabilities. Those suppliers that support the implementation of buying firms' strategic priorities and the generation of competitive advantages can be defined as strategic suppliers[3]. Consequently, a set of strategy-based supplier selection criteria needs to be developed, supporting the creation

[3] In the following, the term 'strategic' supplier is not always explicitly stated if it is clear from the context that not the total supply base is in focus.

of competitive advantages through PSM's strategy implementation support (Culley *et al.*, 1999; Gregory, 1986).

1.2 Research objectives and thesis structure

Having identified suppliers as a potential source of competitive advantages and the need for an alignment of companies' strategic priorities and supplier strategy, the main research objectives of this doctoral thesis are presented. The first research objective is the theory-based analysis of the relationship between companies' strategic priorities and the relevance of selection criteria for strategic suppliers as a pivotal part of PSM strategy development (Krause *et al.*, 2001). The second research objective is the analysis and development of a strategy-based supplier selection approach including organizational, process and content issues. The third research objective is the empirical test of the hypothesized influence of companies' strategic priorities on the relevance of supplier selection criteria and their respective impact on PSM performance measures. The fourth research objective is the empirical verification that PSM performance significantly influences company performance and therefore needs to be recognized as a truly strategic function.

In the following, existing literature will be analyzed to present the existing research findings relevant for this doctoral thesis. Based on the literature analysis results, the research and practice gaps are identified, which represent an important element of the applied research approach and the research question formulation.

Many scholars (e.g. Carter *et al.*, 1996; Farmer, 1978; Nollet *et al.*, 2005; Wisner *et al.*, 2000) have already empirically or conceptually analyzed the strategic role of PSM focusing on the linkage between corporate and functional level. Less examined are the relationships between corporate strategies and specific parts of the PSM strategy such as supplier management, supply markets or internal customers. According to *Narasimhan* and *Carter* (1998), PSM literature lacks empirical evidence that examines the concrete interaction relationship between companies' different strategic priorities and specific parts of PSM strategy. They analyze the personal computer and the automotive industry as examples to explain how differently business unit strategies and environmental factors may influence a company's PSM strategy and its subsequent parts. Their study results show that there is an associative relationship between companies' competitive priorities and their PSM strategies. Based on their results, *Narasimhan* and *Carter* call for a strong linkage of PSM practices to companies' competitive priorities. As further research projects they propose the development of a robust approach which links PSM strategy to different corporate strategies. Only then, PSM managers have a decision framework which supports the implementation of aligned PSM strategies (Narasimhan *et al.*, 1998). In their study, *Narasimhan* and *Carter* have examined relevant links between a company's PSM strategy and specific competitive priorities based on a not further specified collection of corporate strategy elements. They propose the application of other accepted corporate strategy concepts such as *Porter's* (1980) competitive strategies in future research. This doctoral thesis examines the relationship between *Porter's* (1980) widely accepted competitive strategy concept as well as the generally known supply chain strategy approach based on *Fisher* (1997) and the selection priorities for strategic suppliers.

Cousins (2005) recently examined the relationship between main elements of a supply strategy and *Porter's* (1980) two generic strategies cost focus and differentiation focus. He presented a first basic model of how companies' strategic foci influence the nature of strategic supply. In his conclusion, he proposes further research on similar issues with different country foci and a more detailed analysis of established strategy literature. This proposal is followed in a way that the concept of differentiation strategy is analyzed in more detail allowing a distinction between different aspects of differentiation (Fuchs *et al.*, 2000).

Another evidence for the proposed research gap is shown by *Sattler* and *Milling* (2005) who describe the lack of empirical research in supplier management overall despite its high relevance in PSM research and practice. In addition, *Christopher* and *Jüttner* (2000) mention the wealth of conceptual frameworks in strategic supplier management and the corresponding lack of empirical research. Moreover, *Carr* and *Smeltzer* (1999b) as well as *Stuart* (1997) have analyzed in their articles the literature on the empirical linkage between purchasing and strategic management processes. They clearly state the level as unsatisfactory. Research in PSM strategy is stated to be largely independent from research in corporate strategy despite a considerable literature proclaiming the need to regard PSM as a competitive weapon (Carr *et al.*, 1999a; Humphreys *et al.*, 2004; Kannan *et al.*, 2002; Narasimhan *et al.*, 2001a; Nollet *et al.*, 2005; Ramsay, 2001b; Vickery *et al.*, 2003; Vonderembse *et al.*, 1999).

This doctoral thesis aims not only at contributing to PSM research but also to PSM practice development. Consequently, an analysis of PSM practices concerning supplier selection has been carried out to ensure that the intended research efforts also contribute to further supplier selection improvement in PSM practice.

Cousins (2005) describes in his study the situation where PSM managers spent over one year negotiating long-term agreements with suppliers, having already been through an aggressive re-sourcing program. Shortly after the completion of the newly agreed contracts, the suppliers received a letter from the CEO without PSM's prior integration demanding a 10% price reduction across the delivered products. The issue here is obvious: Due to the dramatic financial situation, the company had to follow a cost-focused approach and dramatically reduce costs in order to survive on a short-term. The PSM function instead negotiated long-term contracts that might never be going to be fulfilled if the company could not ensure the short-term survival. This example makes obvious how important it is to link a company's strategic priorities to PSM strategy. PSM's top management needs to be integrated into companies' strategic management processes.

Nowadays, many PSM professionals who are in charge of managing the performance of the supply base do not concentrate much on their companies' strategic priorities and customer expectations (Gaffinen *et al.*, 1994). They focus their processes strongly on the supply side issues, underestimating the critical link between supplier performance and customer satisfaction. This misalignment between customer focus and supply base performance is caused by different aspects. Firstly, there is a lack of understanding of the importance of PSM processes in many companies, resulting in poor support, insufficient staffing, and functional snobbery. Secondly, low information flow, inadequate planning and therefore a lack of customer understanding causes a downward spiral that prevents suppliers from performing up to their actual level of capabilities (Weissman, 2004). This indicates that many companies have not yet put sufficient emphasis on explicitly focusing on the connections between customer requirements, strategic priorities and the way companies are managing their suppliers. Therefore, many PSM organizations still focus on doing tactical deals rather than sourcing from suppliers in such a way that the fulfillments of the customers' needs through top performance are supported (Cox *et al.*, 1995). A recent study by *Gallear* and *Ghobadian* (2004) further indicates a growing awareness of the fact that strategic suppliers have to be evaluated as suppliers of capabilities and not just of products or services. So far, no detailed knowledge enables theory and practice to make specific suggestions on how to select this important part of the supply base.

Tamas (2000) could show in his study that there is neither a strong nor a moderate correlation of PSM's activities with any specific corporate strategy. Seventy percent of the study's respondents agreed that corporate strategy had led to a stronger focus on PSM and supplier management. The results showed also that, in practice, there is still no direct link between companies' strategic priorities and specifically aligned PSM activities. As a result he explicitly calls for further research into how supplier management activities could be better aligned with specific business strategies.

To sum it up, PSM research and practice have been shown to partially lack theory-based concepts and empirical knowledge of the relationships between companies' strategic priorities and the aligned selection of strategic suppliers as an important element of PSM strategy and performance. The basic research questions of this doctoral thesis are therefore as follows:

(a) How are companies' selection criteria for strategic suppliers influenced by their competitive and supply chain priorities?

(b) How are companies' strategic PSM performance levers influenced by their selection criteria for strategic suppliers?

(c) How are companies' financial and market performance influenced by their four strategic PSM performance levers?

The analyses of the potential influence of competitive strategies, supply chain strategies and functional strategies on the selection of strategic suppliers are an important research objective. Furthermore, the influence of supplier selection criteria on strategic PSM performance levers and PSM's impact on company performance is tested to support PSM's increasingly strategic role in companies.

In order to position this doctoral thesis within PSM literature, the major papers for this doctoral thesis focusing on the relationships between supplier selection criteria and/or PSM strategy, companies' strategic priorities and different performance measures are presented. A comprehensive overview on strategic purchasing literature is provided by *Ellram* and *Carr* (1994).

Figure 1 provides an overview of the authors, year of publishing, the main PSM focus of the paper (e.g. supplier selection criteria), the corresponding research focus (e.g. strategic priorities), the applied performance dimension (e.g. company performance), and the main empirical research approach. The overview shows the rich collection of research approaches from single case studies to surveys which are applied in the various papers. From a statistical perspective, the presented quantitative studies have been analyzed with first generation analysis techniques such as correlation and regression analyses except *Cousins'* (2005) work, in which he applies structural equation modeling.

Guinipero and *Brewer* (1993) apply a case study approach to exemplify the influence of supplier selection and evaluation processes on customer satisfaction and costs. *Watts*, *Kim* and

Hahn (1995) provide a framework linking purchasing with corporate competitive strategies. It is their merit to ask for the definition of 'right' in the widely accepted fundamental purpose of purchasing to acquire the right quantity of the right items at the right time and at the right price. They also highlight the importance of consistency between purchasing and the corporate level but also between purchasing and other functions. Additionally, they provide an overview of the paradigm shift in buyer-supplier relationships from adversarial to cooperative and from products to capabilities in supplier selection. *Pearson* and *Ellram* (1995) analyze the differences between small and large companies concerning supplier selection and evaluation. Besides their finding that company size has no significant influence on the relevance of supplier selection criteria, they provide interesting insights about the involvement of various functions in supplier selection and evaluation processes.

Figure 1: Major papers influencing this doctoral thesis

Authors (Year)	Main PSM Focus	Corresponding Research Focus	Applied Performance Dimension	Main Research Approach (Character)
Guinipero/Brewer (1993)	Supplier selection and evaluation process	n/a	Customer satisfaction, costs	Case study (prescriptive)
Watts/Kim/Hahn (1995)	Purchasing strategy (buyer-supplier relationship)	Competitive strategies	n/a	Theoretical development (prescriptive)
Pearson/Ellram (1995)	Supplier selection criteria (large/small companies)	Participating functions in selection process	n/a	Survey, Chi-Square analyses (descriptive)
Narasimhan/Carter (1998)	Sourcing strategies	Competitive priorities	n/a	Survey, correlation analyses (descriptive)
Vonderembse/Tracey (1999)	Supplier selection criteria	n/a	Manufacturing performance, supplier performance	Survey, correlation analyses (descriptive)
Narasimhan/Das (2001)	Strategic purchasing integration and practices	n/a	Manufacturing performance	Survery, regression analyses (explorative)
Kannan/Tan (2002)	Supplier selection criteria	n/a	Company performance	Survey, correlation analyses (descriptive)
Sarkis/Talluri (2002)	Selection criteria process for strategic suppliers	Decision making methods	n/a	Case study (prescriptive)
Kannan/Tan (2003)	Selection criteria for key suppliers (US & Europe)	n/a	Company performance	Survey, correlation analyses (descriptive)
Möller/Törrönen (2003)	Supplier value evaluation	Supplier capability indicators	n/a	Theoretical development (prescriptive)
Ulaga (2003)	Supplier relationship value	Supplier relationship costs	n/a	Grounded theory (descriptive)
Kaufmann (2004)	Strategic supplier relationship BSC	Business unit BSC	n/a	Case study (prescriptive)
Arnold/Eßig (2005)	Supplier lifetime value	Process-oriented SRM	n/a	Theoretical development (prescriptive)
Cousins (2005)	Supply strategy	Cost strategy, differentiation strategy	Business development, market share, relation-ship development	Survey, structural equation modelling (prescriptive)

Narasimhan and *Carter* (1998) correlate in their large-scale study four different competitive priorities and eight sourcing strategies. Their results imply that specific competitive priorities require different sourcing strategies. Unfortunately, they do not apply any performance measure to identify successful strategy linkages. *Vonderembse* and *Tracey* (1999) investigate in their study the extent to which supplier selection criteria and supplier involvement are used by manufacturers. The empirical results support the assumption that the relevance of supplier selection criteria influences performance criteria such as manufacturing and supplier performance. *Narasimhan* and *Das* (2001a) test the influence of purchasing's integration into companies' strategic management processes and purchasing practices on manufacturing performance. Their results emphasize the relevance of purchasing's integration into strategic management processes in order to achieve competitive advantages. It is their merit to highlight the importance of purchasing management topics to elevate purchasing into a strategic function. *Kannan* and *Tan* (2002) identify with their research the relationships between supplier selection criteria and company performance. Their study results indicate that other criteria than cost or quality are most important for company performance. Especially soft, non-quantifiable criteria seem to be relevant although their importance in selection rankings is low. *Sarkis'* and *Talluri's* (2002) research efforts focus on the development of a model for strategic supplier selection highlighting the relevance of supplier selection decisions. They present a criteria weighting approach which supports the structuring of decision problems such as supplier selection. *Kannan* and *Tan* (2003) compare in their study the supplier selection behavior of U.S. American and European companies and identify only minor differences in the relevance of supplier selection criteria. *Möller* and *Törrönen* (2003) address with their approach the problem of evaluating the value creation potential of strategic suppliers. They propose in their paper the application of capabilities as selection criteria for strategic suppliers and highlight the relevance of suppliers' value creation potential as an important part of suppliers' overall performance. *Ulaga* (2003) analyzes in his paper supplier relationship value and supplier relationship costs. Based on in-depth interviews with purchasing managers in manufacturing companies, he identifies eight dimensions in which suppliers can create value. *Kaufmann* (2004b) has developed a X-BSC concept focusing on single buyer-supplier relationships. The X-BSC is an instrument facilitating the implementation of formulated relationship strategies with strategic suppliers. The advantage of X-BSCs lies in the joint agreement on strategic objectives and initiatives between the buying firm and the suppliers supporting a successful strategy implementation in cooperation with strategic suppliers. *Arnold* and *Eßig* (2003) have developed a supplier valuing concept. This approach is based on a supplier lifetime value model which proposes that different economic effects on various impact levels have to be considered when developing an appropriate supplier selection concept. The developed approach consists of the three systems supplier lifetime value, target supplier lifetime value and supplier valuing as integration tool. Although the model provides valuable insights from a theory-based perspective the implementation is difficult due to measurement problems. *Cousins* (2005) has been able to integrate the alignment between a company's strategic priorities and its supply strategy with performance measurement issues. He starts from resource-based view and transaction cost theory to develop a research model that examines the different influences of cost focused and differentiation focused strategies on collaboration types and business outcomes. He concludes that it is really important for PSM managers to understand their companies' strategic priorities in order to accordingly align their PSM strategy. This doctoral thesis represented in figure 1 in the last row builds upon the presented articles to develop a research model that integrates the influence of companies' strategic priorities on supplier selection criteria and their influence on PSM. Moreover, the influence of strategic PSM performance levers on company performance is evaluated to empirically support PSM's claim for a more strategic role.

The research approach followed in this doctoral thesis is influenced by *Ulrich's* (1995) analysis of differences between theoretical research and applied research (figure 2). Theoretical research aims at theory development and theory testing and is driven by the search of truth; applied research in turn wants to construct possible realities and is driven by the usefulness of the developed concepts.

Figure 2: Differences between theoretical and applied research

Characteristics / Research	**Theoretical research**	**Applied research**
Origin of problems	Research	Practice
Kind of problems	Disciplined	Multi-disciplined
Research objectives	Theory development & testing Explanation of existing reality	Construction of possible realities
Kind of statements	Descriptive Free of any value	Normative Assessing
Research regulative	Truth	Usefulness
Progress criteria	Universally application Degree of confirmation Strength of explanation Strength of prediction	Practicability for problem solving

(Source: Adapted from Ulrich, 1995).

Consequently, applied research has to start with practice problems which have not been solved so far. The identified problems then have to be analyzed from a theory-based perspective integrating existing literature and relevant theories (Ulrich, 1995). This procedure ensures a sound theory-based starting point for the detailed analysis of the research objects and the development of useful frameworks. This doctoral thesis is intended rather to follow an applied research approach. Consequently, the presented practice gap supports the identification of the relevant supplier selection problems in practice. Moreover, the presented research gap ensures that the developed model also contributes to the further development of PSM research. The requested integration of problem relevant theories will happen in one of the following chapters.

During recent years, research in PSM has largely increased worldwide but especially in the German-speaking area. This doctoral thesis aims at being a small piece that helps to solve the amazingly large puzzle that PSM researchers and practitioners still have to accomplish. Especially, research in strategic PSM has often been theory-driven, neglecting the empirical proof as requested by *Ulrich* (1995). This doctoral thesis builds up on the existing ideas of strategic PSM; furthermore it examines the already conceptually proposed relationships in literature and finally it empirically tests the hypothesized relationships.

The empirical research approach applied in this doctoral thesis is structural equation modeling (SEM). SEM-based concepts have substantial advantages over first-generation approaches such as principal component analysis, factor analysis or multiple regression analysis because of the greater flexibility a researcher has with the interplay between theory and empirical data. In detail, SEM offers the following possibilities to the researcher (Fornell, 1987):

(a) Modeling relationships among multiple predictor and criteria variables

(b) Constructing unobservable latent variables

(c) Modeling errors in measurements for observed variables

(d) Statistically testing a priori theory and measurement assumptions against empirical data.

Based on these possibilities, *Fassott* (2005) describes four possible test objectives of structural equation models (figure 3).

Firstly, structural equation models can be applied to test whether a theory-based model (1) can be confirmed assuming that the developed model is the correct one. For this reason, co-variance-based analyses such as LISREL (Jöreskog *et al.*, 1996) are preferably applied because they provide the researcher with global goodness-of-fit tests that allow either the confirmation or rejection of the hypothesized research model (theory test). Variance-based SEM approaches such as PLS (Wold, 1975) do not provide global goodness-of-fit criteria. Variance-based SEM approaches rather answer whether a model is more or less predictive but not whether a model can be clearly confirmed or rejected. Although explanatory power measures (2) can be generated for variance-based SEM approaches, co-variance-based SEM approaches are thought to be theoretically superior for this objective (Fassott, 2005). Nevertheless, for both test objectives theory test (1) and high explanatory power (2) research reality shows that this superiority is hardly attainable. For example, sample sizes required to achieve scientifically acceptable significance levels for the acceptance of the null-hypothesis as requested from a theory test are high and difficult to achieve in management practice (Fassott, 2005).

Figure 3: Possible objectives of structural equation models

	Structural Model	**Single Relationships between Constructs**
Significance	(1) Theory Test = **LISREL** more powerful	(3) Hypotheses between two latent Variables = **PLS** more powerful
Relevance	(2) High Explanatory Power = **LISREL** more powerful	(4) Relative Importance/Impact of various exogenous Variables on endogenous Variables = **PLS** more powerful

(Source: Adapted from Fassott, 2005).

Both hypotheses test between two latent variables (3) and the analysis of the relative influences of exogenous variables on an endogenous variable (4) are focused on the specific relationship between latent constructs and their relative impact on endogenous variables compared to others. In this case, the value of each path coefficient is of high interest for the researcher. In these cases variance-based SEM approaches provide more conservative values leading to more accurate conclusions than co-variance-based SEM approaches (Fassott, 2005).

From this point of view, it seems that variance-based SEM approaches are either statistically more conservative (research objectives (3) and (4)) or at least not much inferior to co-variance-based SEM approaches from a research practice perspective (research objectives (1) and (2)). Moreover, variance-based approaches provide further advantages concerning data requirements and construct modeling. Especially the simple construction of formative measurement models provides an opportunity for management researchers to improve the insights of their research for practice.[4]

4 Co-variance-based SEM approaches such as LISREL do also provide the opportunity to construct formative measurement models but their application is less intuitive (Scholderer *et al.*, 2006, p. 65).

As already indicated, a major research objective of this doctoral thesis is the examination of the relative influence of companies' different strategic priorities on the relevance of selection criteria for strategic suppliers as well as their respective relative impact on strategic PSM performance levers and company performance. This puts especially research objective number three into focus. This objective is optimally supported by a variance-based SEM approach such as PLS. Furthermore, a PLS approach supports the usefulness of the research model for practice if it provides accurate statements about the relevance of specific indicators for the latent constructs of the structural model by integrating formative measurement models. As a result, the chosen research approach for the empirical study is PLS.

In the following, the structure of this doctoral thesis is shortly presented (figure 4). It is based on the achievement of three general scientific objectives as underlying sub structure (a similar approach can be found in Gleich, 2001):

(a) Theorieziel:[5] Definition of terms and integration of topic into existing research

(b) Erklärungsziel:[6] Conceptual development and empirical evaluation of relationships between constructs within a research model

(c) Gestaltungsziel:[7] Deduction of recommendations and results for practice and research

Chapter two focuses on the required integration of relevant PSM literature and organization theories. In detail, chapter 2.1 clarifies the relevant terms and definitions. The analysis of PSM in the context of different organization theories is presented in chapter 2.2. Both chapters 2.1 and 2.2 serve as a starting point for the analysis of PSM's role in strategic management in chapter 2.3.

Based on the results in chapter 2, a strategy-based supplier selection approach is developed in chapter 3. Firstly, the relationships between competitive priorities, supply chain priorities as well as functional strategies and selection criteria for strategic suppliers are analyzed in chapter 3.1. Next, a strategy-based supplier selection framework based on existing literature is presented in chapter 3.2. Both chapters 3.1 and 3.2 provide the necessary input in chapter 3.3 for the development of a hypotheses model tested in chapter 4.

In chapter 4.1, the applied survey methodology and relevant descriptive statistics is presented. Furthermore, the formative and reflective measurement models are tested in chapter 4.2.

---

5 Theory objective

6 Explanation objective

7 Configuration objective

Based on analyses in chapter 4.1 and 4.2, the results of the structural model and their implications for PSM research and practice are discussed in chapter 4.3.

Finally, chapter 5 serves as a summary for the most important results and insights of this doctoral thesis.

Figure 4: Doctoral thesis structure

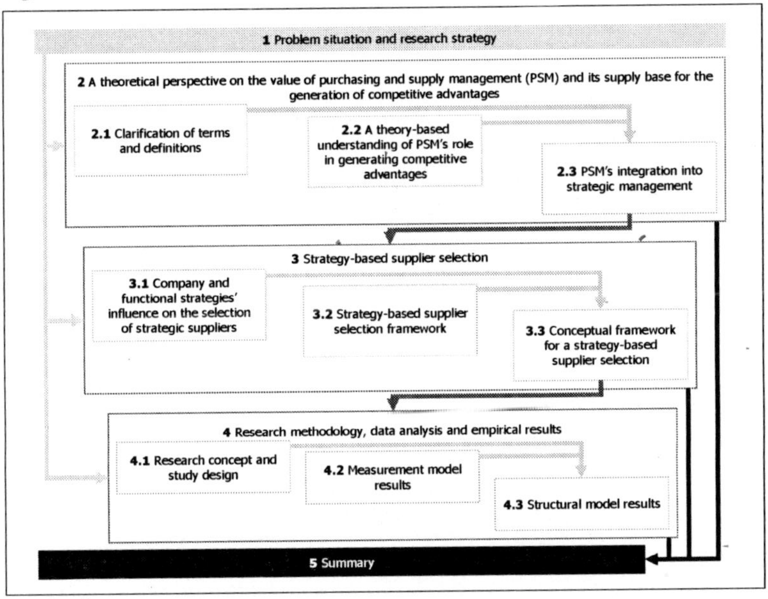

2 A theory-based perspective on the value of purchasing and supply management and its supply base for the generation of competitive advantages

The continuous creation of value is one of companies' most important goals and the only way to survive economically in the long run. A growing number of researchers point to the pivotal importance of business relationships for value creation (e.g. Anderson *et al.*, 1994; Biong *et al.*, 1997; Gemünden *et al.*, 1996; Ravald *et al.*, 1996). Value creation is also regarded as the essential purpose for a buyer and a supplier engaging in any kind of relationship (Sattler *et al.*, 2005; Wilson, 1995). Companies begin to realize that suppliers are creating value not only in the form of cost reductions or quality improvements but also through innovation fostering, risk reduction or increased flexibility (Meier, 1988; van Weele, 1994). Therefore, value creation through suppliers has become an area of interest for many companies. Suppliers can provide buying firms with competitive advantages through their products and services but even more through their many different capabilities (Wagner, 2001) not directly incorporated into their products or services. The identification of those suppliers with the most suitable capabilities is today one of PSM's most important roles.

In the following, a discussion and definition of the most relevant terms and concepts in this doctoral thesis are presented, supporting the theory-based understanding of PSM's ability to create value with strategic suppliers. Therefore, chapter 2.1 is concerned with the proper definition of PSM related terms. Additionally, a definition of strategic value is provided as well as a distinction between resources, competencies and capabilities to support the application of capabilities as selection criteria type in the supplier selection framework which will be developed in chapter 3. In chapter 2.2, different organization theories are analyzed in order to explain PSM's hypothesized impact on the generation of competitive advantages and the existing difficulties in many buyer-supplier relationships discussed in chapter 3 and 4. Additionally, the new St. Gallen management model is applied to PSM in order to provide an integrated framework for the development of a PSM strategy concept. Finally in chapter 2.3, PSM's role in companies' strategic management is described from a process and content perspective in order to develop the necessary understanding for the break-down process from buying firms' corporate strategic priorities to the aligned selection of strategic suppliers.

2.1 Clarification of terms and their definitions

The theory-based concepts in PSM are manifold and rather diverse. Already the discussion about basic terms such as purchasing, procurement, sourcing, supply management or supply chain management can be rather confusing because various researchers and practitioners understand these terms differently. Therefore, the basic PSM related terms of this doctoral thesis will be defined in the following in order to avoid misunderstandings in subsequent sections and chapters. Firstly, the various frameworks of PSM terms will be discussed and definitions for the main elements will be provided. Next, the term value will be analyzed thoroughly from a buyer-supplier perspective. Additionally, a discussion of the term "strategic" shows that this specific term needs to be defined properly to avoid misunderstanding. Thirdly, the concepts of resources, capabilities and core competencies are presented in order to support the reader's understanding of the relevant differences and the reasons for a capability-based supplier selection as proposed in chapter 3.

2.1.1 Purchasing and supply management (PSM)

The basic perception of the business environment heavily influences the understanding and the relevance of buying firms' manifold supplier relationships (Johnsen *et al.*, 2000). It seems necessary to have a detailed understanding of the different terms used in theory and practice to describe the relationships between suppliers and buyers.

There exist many definitions of closely-related terms such as purchasing, procurement, sourcing, supply management, supply chain management, supply network management, supplier management and others. So far theory and practice have not been able to develop a cross-industrially or internationally accepted set of definitions in PSM. *Leenders, Fearon, Flynn* and *Johnson* (2002) state that terms such as purchasing, procurement, sourcing, supply management, supply chain management or logistics are often used almost interchangeably. Until today, there exists no common agreement on the definition of each of these terms.

For the purpose of this doctoral thesis, all relevant terms are discussed and a definition for each is provided. First of all, the terms purchasing, procurement, supply management and sourcing are discussed because they are most closely related to PSM.

The term supply management has been developed in the United States of America; in many European companies the adaptation of this concept is still in its infancy. It has been presented during the early nineties as a new philosophy that seeks to achieve competitive advantages through a strategic view on the value chain and integrate the cooperation efforts of different functions (Gaffinen *et al.*, 1994; Milligan, 1999). The status of supply management in European and especially German companies is quite varying (Hamm, 1998). In short, supply management is as its connotation implies the management of supplies. This focus on the term management is so important because up to now, the management focus in purchasing, procurement and related activities has been strongly neglected for a long time. According to *Bhote* (1989), supply management is not another management fad that will disappear again if ignored. It is also not only a mirage of progress based on the same business-as-usual practices. Even more, it seems to be a new concept in managing the value chain that involves purchasing, engineering, suppliers and others working together to achieve mutual goals.

Another interesting aspect is the change of focus in the evolution from purchasing to supply management. The unit of analysis in supply management has changed from the price and cost perspectives to a value perspective (Matthews *et al.*, 2000; Scheuing, 1998). Supply management acknowledges explicitly that also other aspects than price or total costs matter in buyer-supplier relationships. This implies that supply management can not only happen between PSM managers and suppliers but also requires the integration of other functions such as R&D, marketing or production. As a consequence, PSM managers must understand their internal and the company's external customers in order to create value (Boutellier *et al.*, 1997). In contrast to some authors (e.g. Hollinger, 1992) supply management should not be understood as a concept that only focuses on partnerships with suppliers. Supply management does not automatically imply a change from arm's length relationships to collaborative partnerships.

Jahns (2005a) has presented an integrated set of definitions for purchasing, procurement and supply management which is also used in this doctoral thesis. According to his approach, the terms are defined as follows:

(1) Purchasing

"...Einkauf [Purchasing] steht für die Beschreibung einer rein abwicklungsorientierten, operativen Versorgungsfunktion, die lediglich ausführende Tätigkeiten mit kurzfristigem Wirkungshorizont umfasst (Jahns, 2005a)."[8]

(2) Procurement

"...Beschaffung [Procurement] ist als unternehmensweiter Prozess zu verstehen, der vermehrt den Sicherheitsaspekt und den Kostenaspekt des Einkaufs anspricht. Die Beschaffung beinhaltet einige strategische Fragestellungen und berücksichtigt die technischen und wirtschaftlichen Gesichtspunkte des Beschaffungsmarktes (Jahns, 2005a)."[9]

---

8 "...Purchasing stands for a description of a purely transaction-oriented, operative supply function which only comprises administrative and short-term effect activities."

9 "...Procurement is defined as a company-wide process with a special focus on the security and cost aspects of purchasing. Procurement contains single strategic activities and considers the technical and economical aspects of the supply market."

(3) Supply Management

„Unter dem Begriff des Supply Management [Supply Management] wird die ganzheitliche, integrative beschaffungsseitige Planung, Steuerung und Überwachung der internen und externen Wertschöpfungskette verstanden. Das Supply Management setzt auf strategischer Ebene an, um komplexe Wertschöpfungsfragen aus einer Beschaffungsperspektive zu analysieren, zu gestalten, zu koordinieren und nachhaltig zu optimieren (Effektivitätsziel). Die operativen Einkaufsprozesse können auf dieser strategischen Basis unternehmensintern und unternehmensextern abgewickelt und realisiert werden (Effizienzziel) (Jahns, 2005a).“[10]

Another key term in PSM is sourcing. *Hugos* (2003) states that sourcing includes the activities necessary to acquire the inputs to create products or services (this definition is also applied by van Weele, 2002). According to *Vollmann, Berry* and *Whybark* (1984) sourcing includes activities such as discovery of additional sources of supply, management of the continuity of supply and research supply. Sourcing is different from purchasing, procurement or supply management in such way that the latter terms are often applied to describe PSM's overall evolution level or strategic role within a company (e.g. Freeman *et al.*, 1990; Reck *et al.*, 1988). By contrast, the presented definitions of sourcing and others (e.g. Burt *et al.*, 2002; Lysons *et al.*, 2006) have a clear process view in common. Sourcing is perceived as a process activity which ensures the supply of goods and services building the core process of PSM. The term sourcing is therefore defined as follows:

(4) Sourcing

"Sourcing includes all process activities directly necessary to supply a company with all required products, services and capabilities including activities such as demand identification and bundling, standardization, supply market research, supplier selection or supplier development."

Sourcing is therefore seen as an integrated process from PSM's internal customers to the supply base regardless of PSM's role defined as purchasing, procurement and supply management. Nevertheless, the professional level and content of sourcing activities depends on PSM's understanding as purchasing, procurement or supply management.

As already indicated at the beginning of this doctoral thesis, the term purchasing and supply management (PSM) is applied to cover all clerical and strategic aspects from traditional

10 „The term supply management is defined as the holistic, integrated supply-side planning, management and controlling of the internal and external value chain. Supply management takes a strategic perspective in order to analyze, plan, coordinate and lastingly optimize complex value chains (effectiveness objective). Based on this strategic perspective, operative purchasing processes are planned and implemented within and outside the company (efficiency objective)."

purchasing to strategic supply management and, therefore, includes all aspects necessary for the optimal supply of a company and the generation of competitive advantages with the supply base.

Besides strong internal relationships, this PSM concept cluster also has relevant connections to other terms such as supply chain management and supply network management. In the following, these terms are shortly discussed and supply chain management is defined as a comprehensive term.

The development and perception of concepts such as supply chain management or supply network management have emerged and changed over the last 20 years. In the early 1980s the term supply chain management became known as a concept to manage material flows within an organization (Dubois *et al.*, 2004; Johnsen *et al.*, 2000; Lamming *et al.*, 2000; Vickery *et al.*, 2003). In the early 1990s companies started to extend this concept to their business partners and analyze their business environment from a supply chain perspective. By this time, supply chain collaboration extended only to the closest partners. Second-, third- and nth-tier suppliers and customers were not even identified in many companies. Information from customers had been collected but not shared with upstream partners. Today, supply chain collaboration extends beyond first-tier suppliers and customers and companies are more aware of the complexity of manifold relationships and the increasing uncertainty of business operations. In accordance with this perspective, the developing perception of supply networks formulates the need for more inter-company information sharing and operational coordination with multiple tiers of suppliers and customers in different kinds of relationships (Kemppainen *et al.*, 2003). A supply network can be seen as a set of supply chains of a focal company, including the respective actors, resources and activities (Harland, 1996). Different streams of research can be identified in the recent concept creation of supply networks. For example, the research conducted by the Industrial Marketing and Purchasing (IMP) group (Hakansson, 1982) developed different models to provide a better understanding of business markets and the nature of buyer-supplier relationships. A second approach to supply networks is based on the operations-based and logistics-based research on supply chain management. The relatively new incorporation of the term network into supply chain management seems to make it wider and more strategic by harnessing the potentials of the overall business network of the focal company. This development seems to indicate an increased interest in the competition between company networks rather than solely between companies (Johnsen *et al.*, 2000).

Although many authors (e.g. Cooper *et al.*, 1997; Jahns, 2005a; Kaufmann *et al.*, 2001) foster an integrated conceptualization of the term supply chain management, there exists a school of thought (Christopher, 1994; Göpfert, 1999) strongly focusing on logistics related issues when discussing and defining supply chain management. Although this understanding is rather common in practice, the definitions of the following terms are based on the more integrated understanding from *Jahns* (2005a) and other authors. Following authors such as *Cooper*, *Lambert* and *Pagh* (1997) as well as *Lambert*, *Stock* and *Ellram* (1998) supply chain management is defined as the integration of all business processes from end user through original suppliers.

After having discussed and defined the concepts on a top-level it is necessary to have a detailed look at the concept of supplier management as a core topic of this doctoral thesis (Goffin *et al.*, 1997). According to *Sattler* and *Milling* (2005), supplier management is defined as the management of individual supplier relations. Therefore, supplier management focuses on the upstream part of the supply chain by ensuring the company's supply with the right product at the right time in the right quality at the right place at the right price. *Sattler* and *Milling* claim that supplier management should support production and in consequence business strategy.

This understanding of supplier management represents only a part of the supplier management definition applied in this doctoral thesis. First of all, supplier management might not only include the management of individual supplier relationships but also decisions about the configuration and size of the overall supply base through supplier portfolio management tools (Wagner *et al.*, 2004). Secondly, supplier strategy as part of PSM strategy might not be seen a residuum of production strategy but rather as an integrated part of companies' strategy development and implementation activities. This notion implies that PSM and supplier management not only transform the strategic priorities on corporate level into action on a functional level but also that they can provide relevant input for the strategic management process on corporate level.

From a process perspective, supplier management is often divided into three to four major tasks (Lysons *et al.*, 2006; van Weele, 2002). The first step is supplier selection followed by the evaluation of suppliers' current performance. Depending on the evaluation results, supplier development might be applied as a third step while many practice-oriented concepts such as Siemens' supplier management approach (Siemens, 2005) also include supplier desourcing as a fourth step. In literature, supplier selection is seen as one of the most basic, yet most important PSM activities (Carr *et al.*, 1999b; Correia *et al.*, 2005; Goffin *et al.*, 1997; Krause *et al.*, 2001; Szwejczewski *et al.*, 2001). *González, Quesada* and *Monge* (2004) define supplier selection decisions even as one of the most important activities companies must incorporate into their strategic management processes. The direct and indirect consequences of poor decision making in supplier selection are often critical to the survival of the company because many buying firms have become heavily dependent on their supply base in terms of cost reductions or innovation development. Based on these analysis, supplier management and supplier selection are defined as follows:

(1) Supplier management

"Supplier management includes all activities necessary to manage a company's supply base or individual supplier relationships and select, evaluate, develop or desource specific suppliers in order to generate competitive advantages for the buying firm."

(2) Supplier selection

"Supplier selection is part of supplier management and includes all activities necessary to select a specific supplier for basic materials, products or services on a long-term or short-term basis based on a supplier's respective capabilities and offerings in order to generate competitive advantages."

2.1.2 Strategic value of suppliers

In order to define the strategic value of suppliers a discussion and definition of the terms strategy and value is necessary. At first, the value concept is discussed followed by a short explanation how the term strategic is defined in this doctoral thesis. As a result of these analyses, a theory-based discussion of the term 'strategic supplier' is provided.

The definition of value mainly depends on the perspective taken and is generally defined as something which is synchronous with a perceiver's needs at a specific moment (Cavinato, 1999). The value discussion in business and economics has a long tradition (Payne *et al.*, 2001). Scholars have defined value in a wide variety by focusing on cash benefits (Jackson, 1985), social and financial benefits (Anderson *et al.*, 1999), goal attainment (Woodruf *et al.*, 1996) or competitive advantage (Wilson, 1995). The diversity of views suggests that value may be a multi-dimensional construct that merits multiple measurement approaches. Especially the adaptation of value concepts on costumers and consumers has been supported by many researchers in the marketing field. Much less emphasis has been put on the integration of suppliers in value concepts and the determination by mathematical models. Principally, the value determination concepts for customers can also be transformed into approaches focusing on supplier value (Möller *et al.*, 2003). But as critics of mathematical models of customer value concepts criticize its applicability, the same is true for these supplier value concepts. A literature review (e.g. Al-Mudimigh *et al.*, 2004; Butz *et al.*, 1996; Evans *et al.*, 2001; Flint *et al.*, 2001; Slater, 1997) on value shows that researchers generally conceptualize value in business relationships in two distinct, yet complementary, ways. The first perspective focuses on the worth of a seller's bundle of physical goods and services that is exchanged for some price with the buyer. The second perspective on value focuses rather on the long-term benefits and costs accompanied with a buying firm's relationship to a supplier (Möller *et al.*, 2003). From this perspective, the value of a relationship can be seen as the aggregate worth of all exchanges which will occur between two firms over time (Hogan, 2001). Value seen from a financial perspective (Anderson *et al.*, 1999; Jackson, 1985) can be defined as the net present value of the cash benefits which accrue from current and future transactions. However, these finance-based concepts are limited because they do not account for non-cash benefits such as technology transfer or incremental quality improvements which can be derived from ongoing interactions between companies. Furthermore, they rely on a single discount rate to adjust for the various risks implied in the relevant operations. Unfortunately, the use of a single discount rate often fails to adequately account for the manifold risks inherent in a business endeavor such as a buyer-supplier relationship. The value of a relationship can only be evaluated in the interplay with dynamic market developments (Collis *et al.*, 1995).

In order to overcome this conceptual and practical limitations a strategic perspective (e.g. Woodruf *et al.*, 1996) is applied. According to *Gälweiler* (1987), the decisive criterion to be perceived as strategic is not profits but the development or maintenance of success potentials (this approach is also applied by Large, 2000). These success potentials can be conceptualized as potential competitive advantages which secure the company's long-term survival in its respective markets (Walter *et al.*, 2001). Consequently, it is necessary to take a closer look at the concept of competitive advantages.

Barney (2001) describes three ways of analyzing the concept of competitive advantages. Firstly, competitive advantages can be defined with respect to the actions of current or potential competitors. From this perspective, a company is said to have a competitive advantage when it is engaging in activities increasing its efficiency or effectiveness in such ways that competing companies are not. Secondly, competitive advantages can be seen as a company's ability to generate higher returns than shareholders are expecting. Thirdly, competitive advantages are defined as supernormal profits generated by a company (Shoemaker, 1990). This doctoral thesis follows the third definition of competitive advantages because it provides the researcher with the opportunity to define exactly what kind of supernormal profit the company should generate when following a specific strategy. Furthermore, the concept of supernormal profits and different rent forms is strongly supported by theoretical and empirical research in strategy theory.

According to *Müller-Stewens and Lechner* (2003), competitive advantages are a virtual construct which can be explained by supernormal profits. These kinds of profits are defined as returns which exceed the opportunity costs of the investments in an industry without attracting new competitors. Supernormal profits are not possible under the assumptions of the neo-classical approach of perfect competition. The existence of perfect competition indicates that all companies possess the same homogenous production functions with totally flexible input factors prohibiting supernormal profits (Müller-Stewens *et al.*, 2003; Ramsay, 2001b). In practice, there exist several ways by which companies can generate supernormal profits. Scholars from strategy research have identified four main types of supernormal profits: monopoly rents, Ricardian rents, Schumpeterian rents, and relational rents which are closely related to different strategy theories (Dyer *et al.*, 1998b; Knudsen, 2003; Müller-Stewens *et al.*, 2003; Powell, 2001). These theories will be further analyzed in chapter 2.2.

Following *Gälweiler* (1987), the purpose of companies' strategic management is the development and maintenance of competitive advantages in the form of supernormal profits generated as monopoly, Ricardian, Schumpeterian, or relational rents. From a strategic point of view, value can now be seen as everything which supports the development or maintenance of competitive advantages in the form of the aforementioned rents. In this logic the strategic value of a supplier can be defined as its contribution to the buying firm's achievement of competitive advantages defined as supernormal profits. Therefore, strategic suppliers are defined as those suppliers contributing significantly to the creation of competitive advantages in the form of supernormal profits through their support in strategy development and implementation.

2.1.3 Capabilities, resources and core competencies

Supplier selection approaches traditionally focused on costs, quality and delivery aspects of the supplied products or services. According to *Watts et al.* (1995), a strategy-based supplier selection approach should avoid a short-term, product-based perspective and focus more on suppliers' long-term capabilities. This perspective is also supported by *Burt, Dobler* and *Starling* (2002), *Monczka, Trent* and *Handfield* (2002) as well as *Talluri* and *Narasimhan* (2004) who apply capabilities as analysis level for suppliers. In the following, the terms resources, capabilities and core competences are discussed and an explanation is given why capabilities are the optimal selection level for strategic suppliers from a theory perspective.

Theory and practice have developed varying definitions and interpretations of resources, capabilities, and core competences (Thomsen, 2001). For example, *Barney* (1991) does not make a distinction between resources and capabilities, while *Amit* and *Schoemaker* (1993) as well as *Teece, Pisano* and *Shuen* (1997) make a clear distinction between the two. *Teece, Pisano* and *Shuen* (1997) define resources as strategic assets, such as trade secrets, patents, locations or specialized production facilities, whereas capabilities are defined as the ability to integrate, build, and reconfigure internal and external competencies. *Amit* and *Schoemaker* (1993) also make a clear distinction by defining resources as stocks of available factors that are owned or controlled by the firm, and capabilities as the company's capacity to deploy these resources. Other authors define resources as anything which could be termed a strength or weakness of a given firm – tangible and intangible assets which are semi-permanently tied to the firm. In short, resources are all financial, physical, human, and organizational assets such as people, machines, raw material, knowledge, brand image, or patents (Barney, 1996; Lowson, 2003; Wernerfelt, 1984). Capabilities stand for the ability of the firm to efficiently and effectively combine a set of resources and to attain competitive advantages (Dutta *et al.*, 2005).

After this introduction into the differences of capabilities and resources, the question is whether resources or competencies might be a better analysis level than capabilities. As already defined, strategy is about generating competitive advantages in order to earn supernormal profits in the form of four different rents. In this context, a statement of *Penrose* (1959) seems valuable as she explains that "...a firm may achieve rents not because it has better resources, but rather the firm's distinctive competence involves making better use of its resources...". The term capability is broadly defined and despite many differences in terminology and emphasis, the interpretation of capabilities implies purposeful managerial configuration and orchestration of the company's tangible and intangible resources (Bowman *et al.*, 1998; Howard *et al.*, 1999; Pitt *et al.*, 1999; Porth, 2003). As an additional concept, the term competence has been differently described by a number of scholars (e.g. Snow *et al.*, 1980). *Prahalad* and *Hamel* (1990) define core competence as the collective learning in the organization which creates the ability to consolidate corporate-wide technology and production skills into competencies empowering individual business units to quickly adapt to changing opportunities. Core competences provide potential access to a wide variety of markets. They should make a significant contribution to the perceived customer benefits of the end product and be difficult to imitate (Prahalad *et al.*, 1990). Competencies often focus on technology-based or knowledge-based components. In particular, competencies result from a blending of technology and production skills.

Capabilities represent a complex phenomenon which involves the interactions of individuals and structures and, therefore, are difficult to imitate (Lynskey, 1999; Marino, 1996). Dynamic capabilities (Eisenhardt *et al.*, 2000) are similar to *Amit* and *Shoemaker's* (1993) definition of capabilities and can be defined as a company's ability to integrate, build, and reconfigure internal and external competences to address rapidly changing environments (Halldorsson *et al.*, 2003; Teece *et al.*, 1997). Core capabilities are additionally focused on direct customer value (Stalk *et al.*, 1992; Thomsen, 2001). In this way, competences and capabilities represent two different but complementary dimensions of an emerging paradigm for strategy. Whereas (core) competencies emphasize technological and production expertise at specific points along the value chain, capabilities are more broadly based, encompassing the entire value chain. In this perspective, capabilities are visible to others in a way (core) competencies rarely are (Anonymous, 1992). From this perspective, capabilities seem to be a better approach to develop a new strategic supplier selection approach because they are better measurable and more broadly defined (Sattler *et al.*, 2005).

In general, companies depend on their suppliers' capabilities to manage their day-to-day operations (Loasby, 1998). Even if suppliers' capabilities are effectively packed in the product or service which is supplied so that the buyer does not need to know in detail how it is produced, the buyer still needs to know sufficiently about suppliers' capabilities to judge whether the desired results will indeed be produced (Loasby, 1998) and how flexible the suppliers is from a long-term perspective. Especially for physical goods which are made to order and for services such as marketing or R&D services, it seems obvious that suppliers' capabilities are important to determine their success potential (Large, 2000; Spekman *et al.*, 1998).

As a conclusion, capabilities might include the following characteristics (Croom *et al.*, 1997; Müller-Stewens *et al.*, 2003):

(a) Capabilities coordinate activities of individuals and organizations based on repeatable interaction patterns. They can be perceived as organizational routines that can solve specific problems.

(b) The coordination of activities is rooted in all levels of the organizational culture and comprises cognitive structures, interests and organizational values.

(c) Capabilities contain a potential character since investments in capabilities improve a company's opportunity set in dynamic environments (Loasby, 1998).

(d) The build-up process of capabilities is a complex as well as time and resource consuming task. Capabilities emerge through processes of interaction between individuals, groups and organizations. Therefore, capabilities are difficult to imitate.

Furthermore, a strategy-based, capability-focused supplier selection approach supports a new kind of analysis tools for PSM managers. Whereas a supply chain map, e.g., focuses on the material, money and information flow between the different legal entities, a value chain map turns the attention to the values generated at each stage of the value stream (Cox *et al.*, 1997). In the same way, PSM managers can develop a capability chain map focusing solely on the relevant capabilities down from end consumer requirements to the first raw material suppliers.

2.2 A theory-based understanding of PSM's role in generating competitive advantages

The integration and discussion of organization theories in the context of the analyzed topic of this doctoral thesis is a necessary requirement as discussed in chapter 1.2. PSM's role in generating competitive advantages can be discussed based on a large number of theories. Each theory provides academics and practitioners with valuable insights (Koppelmann, 2002a) and supports the proclamation of cause and effect relationships necessary to build and test hypotheses in management science. Scholars have to select the most appropriate theories to support their research. In the following, strategy theory and new institutional economics theory are analyzed in detail to build up the theory background for a strategy-based supplier selection approach. Elements of these two theories are also proposed by *Cousins* (2005) who argues that any strategic approach to PSM should include resource-based view and transaction cost theory. Therefore, the focus of this doctoral thesis lies on these organization theories.

Moreover, the systems theory based new St. Gallen management model is presented and applied to PSM to support a new PSM strategy approach.

The various strategy theory aspects are analyzed because this doctoral thesis primarily aims at the achievement of competitive advantages through a strategy-based selection of suppliers. They provide the theory basis for the four presented forms of supernormal profits; i.e. monopoly, Ricardian, Schumpeterian and relational rents. Through this analysis, PSM's possibilities to generate competitive advantages are analyzed and discussed.

Transaction cost theory and principal-agent theory are part of the new institutional economics theory and also provide valuable insights into the interaction between buyers and suppliers. Especially the relationship determinants between buyers and suppliers and information asymmetries come into focus (Cousins, 2005). Principal-agent theory explains the increasingly important information situation in buyer-supplier relationships and provides valuable insights into the relevance of true information in supplier selection approaches. Transaction cost theory analyzes the reasons for different kinds of coordination mechanisms relevant for the selection of suppliers. It builds the basis for all kinds of supplier-buyer relationships on the continuum from adversarial relationships to collaborative partnerships.

2.2.1 Strategy theory

Scholars in the strategy field are basically concerned with explaining why some companies are performing better than others. As strategy researchers have searched for sources of competitive advantages, a few important views have emerged regarding the sources of supernormal profits (Amit *et al.*, 1993; Dyer *et al.*, 1998b).

In the following, it will be analyzed how different strategy theories explain to what extent PSM in general as well as strategic supplier selection in particular are related in terms of achieving competitive advantages. As presented, competitive advantages are broadly defined as realized opportunities to generate supernormal profits in the form of monopoly, Ricardian, Schumpeterian or relational rents (Dyer *et al.*, 1998b; Knudsen, 2003; Müller-Stewens *et al.*, 2003; Powell, 2001). Subsequently, four of the most important strategy theories are analyzed as they build the basis for the four rent forms:

(1) Market-based view

(2) Resource-based view

(3) Capability-based view

(4) Relational view

After the discussion of the basic assumptions of each approach, PSM's possibilities to generate different kinds of rents are explored. Based on this analysis, implications for PSM's behavior and supplier selection are discussed. Although some authors (e.g. Sanchez *et al.*, 1997) have criticized the fragmentation character of these strategy approaches, yet they seem valuable when analyzing the influence of supplier selection activities on the generation of supernormal profits.

Ad (1) Market-based view

The roots of the market-based view date back to the Industrial Organization School and *Bain's* (1956) structure-conduct-performance paradigm which explains the success (performance) of a company mainly through industry characteristics (structure) which in turn determine the behavior (conduct) of a company (Jahns, 2003a). Elements of structure include product differentiation, level of industry concentration or cost structures. The term conduct can be characterized by price policy, research & development (R&D) activities, or marketing expenses. The Industrial Organization School model is based on four assumptions:

(a) An above-average performance is the result of a better adaptation to changing conditions of an industry structure

(b) Each company within a specific industry segment has homogenous resources and therefore follows the same strategies

(c) The resources necessary for the implementation of the strategies are completely mobile from a company perspective

(d) Managers decide only rational and in total coherence with the company's interests.

Within this structuralist approach, the challenge for managers is the positioning of the company in the most attractive industry segment because the structure of an industry is more important than the conduct of the managers. The rents generated by this way are usually described as monopoly rents (Fuchs *et al.*, 2000; Müller-Stewens *et al.*, 2003; Spanos *et al.*, 2001).

Based on the Industrial Organization School approach, *Michael Porter* developed his strategic concepts in the 1980s. He introduced the five forces model (Porter, 1985) and the concept of generic strategies (Porter, 1980) proposing that it is not only the industry structure which is important, but also the grounds and nature of competition (Spanos *et al.*, 2001).

Porter's (1985) five forces model is an instrument supporting the analysis of an industry's attractiveness and the respective power relationship between the different actors. It is based on the assumption that five forces and their interaction mainly determine the performance of a company: power of suppliers, power of customers, and threat of new competitors, threat of product or service substitution and rivalry among existing competitors. The industry structure elements strongly dictate the conditions and competitive rules for the focal firm and also the potential strategies available. These strategies are based on *Porter's* (1980) concept of generic strategies consisting of cost leadership strategy, differentiation strategy and focus strategy.

According to *Porter* (1980), every company needs to adopt one of these strategies in order to compete. There appear severe problems for companies engaging in more than one strategic dimension such as cost leadership or brand leadership. This problem is generally referred to as get-stuck-in-the-middle situation (Lowson, 2003; Müller-Stewens *et al.*, 2003).

The presented approach can be used to achieve or protect monopoly rents which are highly dependable on the overall number of companies competing in a specific market. According to *Bain* (1956; 1968) and *Porter* (1980; 1985), there are always potential newcomers trying to enter a specific market as long as there are monopoly rents to achieve. It is vital for the already competing companies to build up entry barriers and prevent potential newcomers to enter their specific market. Three main factors are identified to build up entry barriers (Babu *et al.*, 1989):

(A) Absolute cost advantages

(B) Economies of scale

(C) Product differentiation

In the following, it is analyzed how PSM can support the generation of monopoly rents for the buying firm analyzing the three proposed entry barriers.

Ad (A) Absolute cost advantages

Absolute cost advantages are based on factor cost advantages realized either through lower costs in capital or labor. It is PSM's responsibility to search for suppliers offering the needed product or service (quality) for the lowest total costs on a global scale. Today, the analysis of markets and the identification of those suppliers with the lowest factor costs in low cost countries such as China or India secure the best possible resources and can prevent newcomers from entering the buying firms' respective markets as they have no access to lowest cost suppliers.

Ad (B) Economies of scale

Economies of scale are probably the best known barrier to prevent new companies from entering a market. The concept of economies of scale proposes that a significant increase of the production volume can lead to enormous reductions of total costs through learning-based advantages (Lieberman *et al.*, 1988). With an increased production volume, PSM can increase its market power towards suppliers to negotiate lower prices. But PSM can also act proactively to increase its negotiation power towards suppliers through specification management, volume bundling, increased spend transparency or concentrating the sourcing volume on specific suppliers.

Furthermore, suppliers are proactively positioned to compete against each other in order to maintain a dominant and powerful position for the buying firm. The most suitable supplier will get the contract and be motivated to continuously compete with potential suppliers in order to keep the contract (Knudsen, 2003). From an economies of scale perspective, companies have to become as powerful as possible towards their suppliers to realize cost advantages. This also implies that collaborative approaches with a growing interdependence on specific suppliers need to be avoided at all events. As a consequence, the focus of supplier selection might lie on cost factors and the application of e-procurement tools (de Boer et al., 2001a) to maximize pressure on suppliers by increased spend and supplier transparency.

Ad (C) Product differentiation

Product differentiation can be used to escape price competition and build up entry barriers. This basic principle from *Bain* (1968) had later been applied by *Porter* (1980) to develop his concept of generic strategies (Müller-Stewens *et al.*, 2003). Product differentiation allows companies to differentiate themselves from competitors, e.g., through a strong brand name or advanced technological solutions. Such factors act as entry barriers for potential newcomers and add to the generation of monopoly rents for the established companies. PSM can add to this entry barrier build-up, e.g., by identifying innovations in the supply base to become technology leader or secure the application of ethical guidelines within the supply chain to protect a strong brand name.

From a PSM perspective, the choice of the generic strategy is important in order to optimally manage the supply base and select the most suitable suppliers. Since the market-based view mainly seeks to generate monopoly rents such issues are relevant for the PSM function in order to support companies' positioning efforts. The outside-in perspective of the market-based view provides the PSM function also with requirements such as local content regulations as in the case of the aerospace industry (Moser *et al.*, 2005) or ingredient branding by suppliers as in the case of Intel and computer equipment manufacturers. The supplier selection plays a pivotal role by identifying the necessary supplier capabilities in order to optimally support buying firms' strategic priorities. As a further result from the market-based view, PSM has to integrate outside-in aspects into its supplier selection approach.

In summary, the market-based view provides valuable insights how to generate monopoly rents. Porter's five forces model emphasizes the possible exploitation of market power using deterrent actions as means of achieving competitive advantages. The best way to generate monopoly rents is to deliberately restrict output relative to competitive levels and to include the analysis of behavior and relative market position of competitors (Peteraf, 1993). The decisive step is to relate the focal firm's position to its environment and to maneuver in such a way that it can erect barriers for competitors and potential new entrants such as suppliers to protect its business (Spanos *et al.*, 2001). Having gained an attractive position in an industry, the focal firm can exercise market power towards its supply base and earn monopoly rents.

All these approaches gain competitive advantages by erecting entry barriers, improving the company's power towards both suppliers and buyers, and out-positioning competitors. Therefore, some of the main tasks of the PSM function are the generation of market power towards suppliers and the erection of entry barriers in order to avoid downstream integration. Therefore, the market-based view highlights the importance of building up dominance towards suppliers which needs to be integrated into the prioritization of supplier selection criteria. This aspect needs to be reflected in the optimal selection of strategic suppliers. For instance, PSM can use e-procurement tools as some kind of selection application to create transparency about offerings and prices of suppliers as well as to reduce transaction costs. Other possible instruments of buyer dominance are the use of dual and multiple sourcing strategies.

The market-driven view is still widely held. Nevertheless, many of the aspects of this approach are often rejected in favor of other approaches such as the resource-based view.

Ad (2) Resource-based view

The resource-based view represents a substantial shift towards the individual resources of an organization and draws away the focus from market-based view (Barney, 2001). The resource-based view explains competitive advantages of companies primarily through their internal resources (Duschek, 2004). *Wernerfelt* (1984) built upon the economic theories of

Penrose (1959) and defined companies as a collection of resources, rather than holding purely market positions. From this perspective, companies are no longer seen as homogenous units which have free access to all relevant resources. *Barney* (1996) talks about all financial, physical, human, and organizational assets, the company may gain a sustained competitive advantage from. Therefore, a resource is a basic element a firm *controls* in order to best organize its processes. An employee, a machine, raw material, knowledge, brand image, or a patent can all be considered as examples (Lowson, 2003). Since the end of the 1980s, the resource-based view has been extended to the field of strategic analysis and strategic choice by identifying the importance of resources in strategy development (Jahns, 2003b). The main idea of the resource-based view is to explain success differences between companies based on the different resources a company has access to (Prahalad *et al.*, 1990). These success differences are only explainable if companies have limited access to heterogeneous resources. The essence of the resource-based view is therefore its focus on the individual resources of the organization; rather than a market-based view proposing commonalities with other companies within industries. Understanding the particular sector is important but organizations should seek their own individual solutions in this context. Sustainable advantages come from the exploitation of unique resources of individual organizations which can be seen as bundles or clusters of resources (Foss, 1999). These clusters have to be managed and combined to create the differences supporting a differentiated strategic positioning. However, they can not easily be re-arranged to take account of market opportunities. Accordingly, organizations must define opportunities in terms of existing resources and focus on unique expertise.

The resource-based view defines success as the ability to generate competitive advantages (Carr *et al.*, 2002). These advantages are somewhat based on monopoly rents[11] but mainly on Ricardian rents. Ricardian rents can be achieved by owning or controlling scarce resources and assets which stands in contradiction to the neo-classical assumptions of a perfect market (Müller-Stewens *et al.*, 2003). Ricardian rents are therefore based on strategically important but restricted resources within an industry sector. Companies possessing or controlling such resources have lower average costs than their competitors leading to Ricardian rents as long as their competitive advantage is sustainable (Spanos *et al.*, 2001). Beside the heterogeneity of the resources, there are other requirements to generate Ricardian rents in a sustainable way (Hoyt *et al.*, 2000; Peteraf, 1993; Ramsay, 2001b). As a first condition, it is necessary that there exist *ex ante* restrictions of the competition on strategically valuable resources, because

[11] From a resource-based view, monopoly rents can be generated by the possession of unique resources such as monopolistic-like concessions for postal services or unique location advantages for a restaurant.

otherwise, the price of these resources would be so high that there wouldn't be any possibility of generating supernormal profits (Jap, 2001). This implies the image of a market for strategically valuable resources where companies can get these valuable resources in two different ways: Either based on luck or based on managerial foresight acquiring strategically valuable resources cheaply before others see the strategic importance of a specific resource. Consequently, the competition on strategically valuable resources is already decided when the acquisition of the resources takes place. Resource-picking and resource-controlling are the most important and decisive characteristics of managers. As a second condition, strategically valuable resources should not be totally mobile or tradable. Otherwise, prices for these resources would be so high that there will be no possibility to achieve supernormal profits. As a third condition, there must exist *ex post* restrictions on the competition to ensure the heterogeneity of the resources through limited imitation and substitution possibilities (e.g. specific property rights, historic unique starting conditions, information asymmetry, etc.). If resources are imitable or substitutable competitive advantages of formerly valuable resources could be neutralized by competitors (Fuchs *et al.*, 2000; Knudsen, 2003; Müller-Stewens *et al.*, 2003).

Criticism about the resource-based view focuses especially on the problem that in practice strategically valuable resources are hardly identified *ex ante,* leaving managers alone with this decisive question as well as the missing market perspective. In addition, the resource-based view seems to be a too static perspective in an ever changing business environment (Priem *et al.*, 2001). The *ex ante* problem and the missing market perspective are closely related since *Barney* (2001) defines the value of a resource in the context of current market conditions. Taking this into account, the static perspective might also be overcome and turned into a dynamic approach which enables managers to react to or proactively act in advance to changes in the market environment.

Instead of focusing on transparency, buyer dominance, or market-driven requirements when searching for monopoly rents, generating Ricardian rents in PSM emphasizes value maximization through acquiring or controlling strategically valuable resources. The role of supplier management lies primarily in the identification and acquisition of strategically valuable supplier resources. In some cases, companies may use strategic alliances to obtain resources such as technologies, locations or market data possessed by other companies and enabling the generation of competitive advantages. However, only specific supplier relationships can generate Ricardian rents. Arm's-length relationships which are often the case when buying in an open marketplace with theoretically perfect competition are categorized by non-specific asset investments, minimal information exchange, separable technological and functional systems and low transaction costs with minimal investment in governance mechanisms (Badaracco, 1991). These open markets provide supplier assets available to virtually every company. Therefore, resources and assets are available for anyone and hardly generate Ricardian rents as it conflicts with the requirements stated above. The only way to achieve a competitive advantage in this case is to increase the bargaining power relative to the suppliers resulting in monopoly rents. Generating Ricardian rents means rather anticipating industry trends, identifying potentially valuable resources in the supply base and finally acquiring or controlling the identified supplier resources before other companies.

In summary, PSM can support the achievement of Ricardian rents by identifying strategically valuable resources such as patents, product development knowledge, relevant market information or market access possibilities in the supply base. After the identification and acquisition of these resources, PSM managers have to decide about the kind of protection against imitation and resource mobility, e.g. through detailed, long-term supply contracts or the build-up of idiosyncratic relationships.

Ad (3) Capability-based view

The capability-based view is closely related to the resource-based view but there are also important differences. The most important distinction is the difference between the points in time when the specific Ricardian (resource-based view) or Schumpeterian[12] (capability-based view) rents are generated. While the resource-based approach generates Ricardian rents until the point in time when a company gets an underestimated resource under control, the capability-based view assumes that the Schumpeterian rent generation starts at that point in time when a resource is under control of the company. As the resource-based view claims that the control of a strategically valuable resource should 'automatically' lead to the generation of Ricardian rents, the capability-based view assumes that only the clever coordination and combination of resources lead to competitive advantages in the form of Schumpeterian rents. Therefore, rent generation mechanisms are totally different. While the resource-based view is focused on superior information and the identification of underestimated strategically valuable resources, mechanisms of the capability-based approach concentrate on the clever coordination and use of different resources (Müller-Stewens *et al.*, 2003).

---

12 The Schumpeterian (entrepreneurial) rents are based on the Austrian school whose most prominent scholar is Schumpeter. He introduced the term creative destruction meaning that rents will become available for entrepreneurs who introduce new technology, explore new markets and search for new ways of creating value (Jacobson, 1992, p. 787).

Capabilities reflect organizations' ability to use their competencies as well as combine and coordinate resources. It should be noted, however, that the strategic value of resources, capabilities and competencies is dynamic and constantly changing (Teece *et al.*, 1997). This implicates that capabilities refer to a firm's capacity to deploy resources or competencies, usually in combination, using organizational processes for affecting a desired result (Amit *et al.*, 1993). Capabilities are also seen as complex interaction, coordination and problem solving patterns of an organization. They are often related to specific groups or members of an organization and are built-up in long-term development process. Therefore, capabilities are less easy to substitute or imitate than specific resources. As a result, *Teece*, *Pisano* and *Shuen* (1997) define capabilities as companies' ability to integrate, build, and reconfigure internal and *external* competences to address rapidly changing environments. Therefore the central strategic task is the management of the internal and *external* resources as well as competencies using capabilities to solve organizational problems.

In summary, the capability-based approach explains competitive advantages based on the heterogeneity of companies but, in contrast to the resource-based view, heterogeneity is not a result of firms' different resources but of firms' different capabilities using the same resources differently. Concerning the generated rents, the capability-based approach is based on Schumpeterian rents which result from dynamic and clever managerial and entrepreneurial decisions about the right combination and coordination of resources.

An important distinction between the concepts of monopoly and the Ricardian rents on the one hand and the concept of Schumpeterian rents on the other hand involves the existence of *sustained* competitive advantage (Knudsen, 2003). According to the Schumpeterian concept, there exists no such thing as *sustained* competitive advantages but only *temporary* competitive advantages (Jacobson, 1992). The nature of competition suggests that no replicable strategy will allow businesses to earn supernormal profits in the long-term and thus *Porter's* (1980) three generic strategies cannot be enough to gain a sustained competitive advantage (Jacobson, 1992). It is argued that in order to be competitive in the long run, a company must string together a chain of unsustainable competitive advantages. Furthermore, greater returns often flow to companies which destruct their own playing fields and change the established rules of competition. These companies can destroy the competencies of industry leaders by changing the industry's critical success factors and make the leader's competencies obsolete using new technology or other innovations to establish a superior value-creating process (Müller-Stewens *et al.*, 2003). Schumpeterian rents diminish therefore as innovations become commonplace. The only way to stay ahead of competition is to be innovative. The innovation process is a critical strategic process central to the development of competitive advantages from a Schumpeterian perspective. Innovations are everything which improves the way something is currently being done to meet customer needs more efficiently, more effectively or within a better risk-opportunity situation (Knudsen, 2003).

The implications of the capability-based view for PSM are quite different from those of the resource-based view. Finding suppliers with innovative products or services becomes more and more vital for the survival of the buying firm. It is also important to keep speed high as competition is always present. In practice, supplier selection approaches explicitly taking such aspects into consideration are still quite seldom. Very few, if any, supplier selection approaches can be said to have their roots in Schumpeterian discoveries. However, in recent years the inherent nature of competition has changed, relying more on constant technological innovation and rapid entry into new markets. Thus, the need for supplier selection concepts involving these issues becomes more urgent than ever before.

Recently, PSM practice has begun to realize the importance of scanning the supply base for potential partners and innovations, providing buying firms with new growth opportunities or early knowledge about attractive innovations, e.g., in the form of new materials, services or customer knowledge. This often allows the realization of first mover advantages when entering into new customer, product or geographical markets (Lambe *et al.*, 1997). The consequence for PSM is the necessity to scan the supply base for innovations which can be used to improve buying firms' own customer offerings. By being closer to the end consumer, it is often possible for buying firms to better assess innovations from a consumer perspective and thus realize the potential value of suppliers' innovations. Even though some profit opportunities are uncovered by pure chance, successful companies often have more information than others through systematic search processes. This knowledge gives them an advantage in ascertaining market inefficiencies.

It is very important to realize at this point that innovations come from finding and integrating *ex ante* opportunities in the supply market that possess *ex post* value in the customer market (Peteraf, 1993). As a result, the existence of true Schumpeterian rents depends on the possession of superior information and management capabilities. The entrepreneurial role of PSM is to gather information, evaluate possibilities and realize opportunities (Jacobson, 1992).

A last important factor is the integration of suppliers into joint efforts to develop new products or service offerings. PSM managers must make use of suppliers' new ideas and innovative capabilities to add value to new business solutions either through joint efforts or 'black box' supplies (Bozdogan *et al.*, 1998; Fraser *et al.*, 2003). From a Schumpeterian perspective, PSM managers have to combine suppliers' capabilities and buying firms' capabilities to develop new customer market solutions.

Ad (4) Relational view

Recently, the market-based view, the resource-based view and the capability-based view were supplemented by the relational approach developed by scholars such as *Dyer* and *Singh* (1998b) or *Smith, Carroll* and *Ashford* (1991) as another independent approach in strategy theory (Contractor *et al.*, 2002; Kaufman *et al.*, 2000). In contrary to the market-based approach which suggests that supernormal profits are primarily a function of a company's membership in an industry with favorable structural characteristics or the resourced-based view proposing that supernormal profits are due to the resource heterogeneity of companies the relational view highlights that relationships themselves are a possibility to create competitive advantages (figure 5) (Müller-Stewens *et al.*, 2003).

Compared to the capability-based view, collaborative capabilities supporting the development of new solutions are not sufficient to generate relational rents. In fact, two or more companies have to create value which otherwise could not be created by a single company to generate competitive advantages in the form of relational rents (Carr *et al.*, 2002; Zajac *et al.*, 1993). An increasing number of researchers propose that firms' competitive advantages reside not simply within the boundaries of what the company owns and controls, but also in the idiosyncratic interfaces it develops with other companies such as suppliers (Chung *et al.*, 2004; Duschek, 2004; Dyer *et al.*, 1998b). There is a growing recognition that competitive advantages can not only lie *within* a company but also *in* relationships *between* suppliers and buyers (Jap, 2001). A firm's competitive position can then be defined as control of resources as well as access to resources controlled by other companies such as suppliers (Araujo *et al.*, 1999). Proponents of the relational view emphasize that companies' strategic resources may extend beyond their boundaries. This indicates that companies investing into specific relationships may realize an advantage over competing companies which are unable or unwilling to do so. Thus, idiosyncratic intercompany linkages may be a source of relational rents. This analysis suggests that companies' strategic resources may span company boundaries and may be embedded in intercompany routines and processes. Therefore, the relational view focuses on dyadic and network relations as unit of analysis (Dyer *et al.*, 1998b).

Figure 5: Comparison of market-based, resource-based and relational view

Dimensions	Market-based View	Resource-based View	Relational View
Unit of analysis	Industry	Firm	Pair of network of firms
Primary source of supernormal profit returns	• Relative bargaining power • Collusion	• Scarce physical resources • Human resources/know-how • Technological resources (e.g. process technology) • Financial resources • Intangible resources (e.g. reputation)	• Relation specific investements • Interfirm knowledge sharing routines • Complementary resource endowments • Effective governance
Mechanisms that preserve supernormal profit returns	• Industry barriers to entry: -Government regulations -Production economies / sunk costs	• Firm-level barriers to imitation: -Resource scarcity / property rights -Causal ambiguity -Time compression diseconomies -Asset stock interconnectedness	• Dyadic / network barriers to imitation: -Causal ambiguity / resource indivisibility -Time compression diseconomies -Interorganizational asset stock interconnectedness -Partner scarcity -Institutional environment
Ownership / control of rent-generating process / resources	• Collective (with competitors)	• Individual firm	• Collective (with trading partners)

(Source: Adapted from Dyer *et al.*, 1998b).

Idiosyncratic intercompany relationships need to be developed in order to generate relational rents. This can happen in the form of investments in relation-specific assets, cultural understanding, substantial knowledge exchange and joint learning. Moreover, the combination of complementary, but scarce resources or capabilities which result in the joint creation of unique new products, lower transaction costs and more effective governance mechanisms are means to build up intercompany relationships (Dyer *et al.*, 1998b; Kaufman *et al.*, 2000).

Relation-specific assets can be divided into site specificity, physical asset specificity and human asset specificity (Cousins, 2005). For example, human asset specificity exists from a PSM perspective if a buyer and a supplier gain mutual experiences in specific production stages and thereby establish a common language, knowledge, and routines which represent more efficient communication structures than between other companies. Site specificity can be developed by a sequenced value chain which is organized within a network placed spatially close to each other such as the Smart lean manufacturing system in Hambach. Physical asset specificity finally is achieved by the investment of partners in co-specialized equipment (Duschek, 2004).

Knowledge exchange and joint learning mainly concern the problem solving capacities of cooperating companies necessary for the generation of competitive advantages. For example, many suppliers present to buyers an important source of unique ideas which can result in product or process innovations (Powell *et al.*, 1996).

Complementary resources and capabilities represent a source of relational rent through the option of mutual expansion. From this point of view, it is best if none of the involved companies has access to similar capabilities although a minimum organizational and cultural fit between the companies are prerequisites. Therefore, buying firms have to identify and evaluate potential complementarities in their supply base (Chung *et al.*, 2000). As an example, the low budget chain store H&M worked together with the star designer Carl Lagerfeld in order to provide its customers with brand fashion at moderate costs successfully combining the capabilities of both organizations (Kappeler, 2004).

Finally, governance structures represent an important element of relational competitive advantages. Protection against opportunism is seen as an important component of an optimal network management due to the specificity of the resources and capabilities. High co-specialization reduces the value of these resources and capabilities in alternative uses. Therefore, there exists a risk of opportunistic behavior which needs to be optimized in order to engage in joint value creation processes (Zajac *et al.*, 1993). For instance, suppliers could apply governance structures based on legal settlements or third-party enforcements of agreements in case of conflicts.

There would be no possibilities of generating long-term, supernormal profits through relational competitive advantages if there weren't any mechanisms preserving relational rents. For example, mechanisms such as causal ambiguity and time compression diseconomies preserve relational rents (Hoopes *et al.*, 2003). The development of trust, for instance, is subject to causal ambiguity because it is a complex and situation-specific process. Moreover, the development of trust or partner-specific absorptive capacity is subject to time compression diseconomies because it cannot be developed quickly nor can it be bought in a marketplace (Jahns *et al.*, 2005c).

Additionally, relational rents can be preserved, for example, through interorganizational asset interconnectedness, partner scarcity or specific institutional environments (Dyer *et al.*, 1998b). Interorganizational asset interconnectedness occurs in cumulative increments on an existing stock of assets held by a company or its alliance partners. For example, if a supplier builds its new plant next to the buying firm's facilities, the supplier is willing to make a site-specific investment. Once this site-specific investment is made by the supplier, the supplier and the buying firm are together investing in an economical conveyor belt that is a highly specific asset and not in a supplier paid truck transportation system, a general-purpose asset, in order to organize the transportation from the supplier to the buyer. This example shows how initial relation-specific investments create conditions which render subsequent specialized investments economically interesting. The key strategic implication of this preventing mechanism is that partnerships between suppliers and buying firms may need to make bundles of related relation-specific investments to realize the full potential of a relationship (Duschek, 2004).

Partner scarcity exists if companies do not find partners with complementary strategic resources and relational capabilities. This is a particular problem for late movers who enter, for example, markets where few companies with local knowledge, contacts and distribution networks exist. A good illustration is the problem of Toyota when it was entering the U.S. market and found no U.S. suppliers willing to work in Toyota's preferred partnership fashion (Duschek, 2004).

Companies can also combine resources or jointly developed capabilities in such a way that the resulting resources are both idiosyncratic and indivisible. A good example for this case is the VISA organization which is actually an asset jointly created by about 23'000 banks. The VISA brand name and distribution network are idiosyncratic and indivisible assets which are collectively owned by the participating banks in a large multi-company relationship (Dyer *et al.*, 1998b).

Specific institutional environments can also be seen as preventive mechanisms because trust fostering, institutional environments support the creation of relational rents better than others. The trust-based Japanese culture with its specific interaction mechanisms may build a better environment to prevent relational rents than others (Dyer *et al.*, 1998b).

By moving away from arm's-length exchanges and specializing their relationships through idiosyncratic investments, knowledge exchange, complementary competencies, and more effective governance mechanisms, buying firms and their suppliers can create the potential for earning competitive advantages (Jap, 2001).

As a consequence, PSM managers might create relational rents based on close interactions with suppliers. The other three presented forms of rent generation mechanisms can theoretically be generated by adversarial or coordinative relationships. Relational rents are the only form completely relying on collaborative partnerships (figure 6).

Figure 6: Summary of basic rent possibilities and consequences for PSM

Type of rent	Monopoly rents	Ricardian rents	Schumpeterian rents	Relational rents
How to gain (sustainable) competitive advantage	Exploitation of market power using deterrent actions such as erecting barriers for newcomers, enhancing your power towards both buyers and suppliers, and outmanoeuvre competitors.	Develop and maintain resources and capabilities that are valuable, rare, costly to imitate and non substitutable.	Act quickly and seize opportunities based on the discovery and possession of superior information and possibilities regarding new innovations. The entrepreneurial role is to gather information, evaluate possibilities and realize opportunities.	Build up idiosyncratic inter-company linkages through relation-specific invest-elments and the combination of resources in unique ways.
PSM's contribution to rent generation	Maintaining or enhan-cing power over suppliers, and refrain from long-term relationships as this might reduce power advantage. Reduce cost in all stages of the purchasing process.	Finding and matching suppliers with resources complementary to one's own resources. Development of close supplier relationships that become idiosyncratic, valuable and hard to imitate.	Skimming and scanning, finding and materializing on new suppliers and new, innovative products and services, being alert and ready to act quickly.	Building an image of trust for the buying firm that enhances the creation of collaborative relationships. Fostering of information exchange. Identifying complementary resources.

(Source: Adapted from Dyer *et al.*, 1998b; Knudsen, 2003).

In summary, figure 6 shows the different possibilities to generate competitive advantages in the form of monopoly, Ricardian, Schumpeterian and relational rents and PSM's specific contributions.

The four strategy theories and the respective rent concepts have been analyzed from a PSM perspective showing how PSM can have a truly strategic influence for the generation of competitive advantages when integrating these insights into its supplier selection activities. These findings will support the hypothesized relationships between the selection criteria for strategic suppliers and strategic PSM performance levers in order to create competitive advantages in the empirically tested research model in chapter 4.

2.2.2 New institutional economics

The research stream of new institutional economics focuses on the analysis of institutions such as companies, markets or political systems. The key ideas from new institutional economics approaches provide practitioners and researches with insights about the appropriateness of alternative organization forms such as markets, hierarchies or networks for different exchange situations and the relevance of information asymmetries (Müller-Stewens *et al.*, 2003). In the context of buyer-supplier relationships principal agent theory and transaction cost theory are relevant. According to *Cousins* (2005), transaction cost theory is especially important for the understanding of strategic PSM. The principal agent theory and the transaction cost theory are discussed in detail in the following.

Principal-agent theory is directed at the relationship between a principal and its agents in which one party (the principal) delegates an assignment to another (the agents). From a PSM perspective, the principal is the buyer whereas the supplier has the function of the agent. Principal-agent theory is concerned with resolving two problems which can also occur in buyer-supplier relationships. The first is the agency problem arising when (a) the objectives of the principal and the agent conflict and (b) it is difficult or expensive for the principal to verify what the agent is actually doing. Firstly, there exists an objective asymmetry between the buyer and the supplier because the buying firm is interested in low prices and good quality whereas the supplier aims at high margins at an acceptable quality (Kaufmann, 2001). Secondly, the problem is that the principal cannot verify whether the agent behaved appropriately. As a result, there appear information asymmetries between the buyer and the supplier in form of hidden intentions, hidden characteristics and hidden actions (figure 7) (Aberle *et al.*, 2002).

Figure 7: Basic types of behavior insecurities

		Behavior after the contract agreement is...	
		... observable	**... not observable**
Behavior before contract agreement is...	**... determined**	**Hidden Characteristics** (Quality Insecurity)	Not discussed
	... not determined	**Hidden Intention** (Hold up)	**Hidden Action** (Moral Hazard)

(Source: Adapted from Eßig, 1999).

The possible information asymmetries are described in the following. Hidden characteristics are related to behavior patterns where the agent's behavior is already determined before a contract is agreed and the real performance becomes known to the business partners *ex post*. In this case we talk about the real performance levels of buyers and suppliers. For instance, if buyers know the real performance levels of their suppliers in terms of cost, quality, flexibility or innovation only *ex post* buyers should improve their selection processes by establishing reliable indicators for the performance of suppliers. Furthermore, buyers can negotiate a flexible contract agreement allowing the adaptation of their payments according to supplier's real performance as measured *ex post* (Eßig, 1999).

Hidden intention occurs, for instance, if suppliers make opportunistically use of existing performance specification gaps in the contract for their own profit increasing buyers' purchasing costs. Three cases of not determined but observable behavior can be distinguished. First, incomplete contracts allow suppliers to change their performance to a certain degree without legal consequences. A second case is the investment in transaction-specific assets. For example, if a company buys specific software from an IT-supplier and this supplier stops the further development of the software the buyer is under pressure either to change the supplier or to pay a higher license fee for the supplier's further investments into IT developments. Finally, a hold-up situation evolves if suppliers act intentionally and misuse their dominant position and the trust of buyers. In order to prevent such a scenario buyers must take appropriate actions which may include the build-up and selection of second suppliers (Koppelmann, 2002b).

Hidden action is concerned with a behavior that is not determined *ex ante* and not observable *ex post*. In contrast to the hidden intention situation, buyers cannot observe suppliers' behavior *ex post* but only the results of the actions. The results are not only based on suppliers' specific activities but also on external influences (Eßig, 1999) such as suppliers' suppliers. Therefore, it is difficult to assign the success or failure of a project directly to a specific suppliers' real performance. PSM managers might apply risk-rent sharing agreements in their selection process to motivate suppliers and overcome the hidden action problem (Jahns *et al.*, 2005c).

The second premise is the risk sharing situation which arises if the principal and the agent prefer different actions because of different risk preferences (Trienekens *et al.*, 2001). The principal is assumed to be risk neutral while the agent is risk avers (Kaufmann, 2001). This leads to the presumption that the buying firm is risk neutral while the supplier is supposed to be risk avers. This situation does not apply to all buyer-supplier relationships because suppliers are often much larger than buyers; therefore they might be able to bear more risk than the buyers. Such considerations need to be integrated into supplier selection processes through specific criteria.

In summary, principal-agent theory provides useful insights for PSM managers to optimize their supplier selection and performance management. The example of hidden characteristics shows the importance of reliable indicators. Hidden intentions exemplify the importance of trust and independence considerations as selection criteria and the concept of hidden actions in buyer-supplier relationships encourages PSM managers to transform motivation instruments such as risk-rent sharing agreements from other management areas into their selection process.

The unit of analysis of transaction cost theory (Smeltzer *et al.*, 1998) is the organizational form of exchange relationships between specialized entities in an economic system (Coase, 1937; cited in Müller-Stewens *et al.*, 2003). Transaction cost theory neglects material and manufacturing costs and concentrates on the costs arising in order to make a transaction happen (Heide *et al.*, 1995; Wohlgemuth, 2002). These costs consist mainly of search costs, negotiation costs, processing costs, control costs and adaptation costs (Monczka *et al.*, 1995; Picot, 1982, 1991) which occur through the coordination of economic activities between different entities.

Transaction cost theory's main premise is that the optimal form of organizational coordination should be selected based on lowest transaction costs. The core of the transaction cost theory is therefore a total transaction cost comparison between different organizational concepts mostly described as hierarchies, markets and networks or hybrid organizations (Aberle *et al.*, 2002; Müller, 2003). Figure 8 shows that transaction costs mainly depend on basic assumptions about transaction partners' behavior, transaction characteristics and conditions of the transaction atmosphere.

Figure 8: A basic model of transaction cost theory

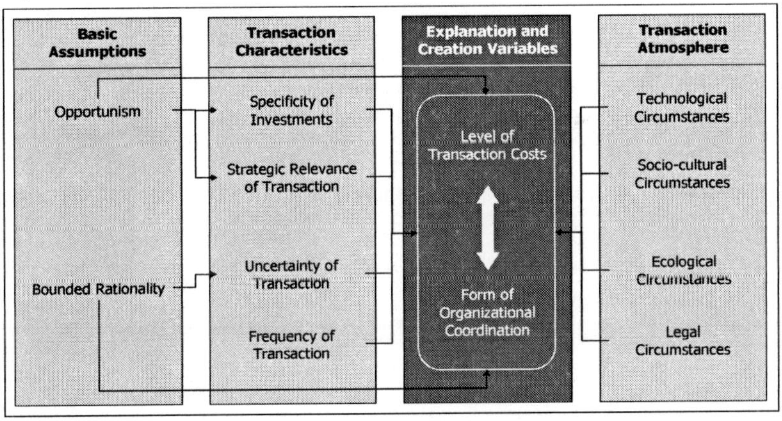

(Source: Adapted from Wohlgemuth, 2002).

Transaction cost theory assumes that managers (agents) have bounded rationalities (Simon, 1957) due to their non-perfect information status and bounded information processing capabilities (Cousins, 2005). Furthermore, agents are seen as individuals acting opportunistically in their own interest and lying or deceiving in order to achieve their goals. Transaction cost theory therefore proposes that, in absence of some form of governance mechanism, agreements between organizations will always be subject to risk from opportunistic behavior (Hoyt *et al.*, 2000).

Transaction characteristics in the form of specificity of investments, strategic importance, uncertainty and frequency of the transaction mainly influence the level of transaction costs and the appropriate form of organizational coordination. The specificity of investments is seen as the most important transaction cost driver (Correia *et al.*, 2005) and represents also a relevant supplier selection criteria. The specific investments can be divided into physical assets, human assets and site assets and develop a dependency or interdependency between the exchange partners (Cousins, 2005). Another important driver of transaction costs is the strategic relevance of the coordinated activities. Especially if the survival of the entity depends on the specific activity, hierarchies are preferred to market coordination. Thirdly, uncertainty is a measurement for possible changes such as economic developments or customer demands. In supplier selection, this aspect is often reflected in different forms of flexibility criteria. More uncertainty leads to more transaction costs because more information has to be gathered in order to conduct the necessary transactions. The higher the uncertainty, the more efficient seem hierarchical organization forms. The frequency aspect is often only used in combination with other determinants to make decisions about the optimal form or organizational coordination (Müller-Stewens *et al.*, 2003).

The transaction atmosphere also determines the level of transaction costs and therefore the optimal form of coordination. It contains all technological, socio-cultural, ecological and legal circumstances such as corporate taxes, communication costs or adherence to formal hierarchies which have an impact on the transaction costs between the different entities of an economic system (Picot *et al.*, 1997; Wohlgemuth, 2002). These aspects are especially interesting for international companies because it shows that the relevance of different supplier selection criteria is also subject to local conditions.

From a PSM perspective, transaction cost theory proposes that a buyer and a supplier will develop an ongoing relationship when transaction costs associated with maintaining the relationship are less than the transaction costs of buying in an open market or formally integrating the supplier into the buying firm (Monczka *et al.*, 1995). These considerations need to be taken into account when selecting strategic suppliers, e.g. through the prioritization of suppliers' collaborative capabilities. PSM seeks to minimize the costs related to searching, evaluating, selecting, and developing suppliers' capabilities. The transaction costs of the repeated search, evaluation, selection and development of suppliers from an open market are often higher than the transaction costs of ongoing relationship with a few specific suppliers (Grundlach *et al.*, 1993). This implicates that the common imperative to always buy non-strategic commodities from changing suppliers at arm's-length relationships might not be as optimal as assumed according to simple product price comparisons (Dyer *et al.*, 1998a). The decisive criterion for the optimal organizational form should always be based on the rent generation possibilities as discussed in chapter 2.2. For instance, it might make sense for one company to buy non-strategic items from changing suppliers leveraging its market power and realizing monopoly rents while another company engages in joint process improvements with a single supplier in order to generate relational rents. The latter approach often requires transaction-specific investments. These investments are assets that are uniquely tailored to a particular exchange relationship and have a low value outside the focal relationship (Kleinau, 1994). In industrial buyer-supplier relationships, buying firms may make investments in tooling, equipment and organizational procedures which are uniquely tailored to the specific relationship with an individual supplier. For example, Xerox incorporates supplier-designed components into many of its products which require idiosyncratic adaptations of production lines and procedures to individual suppliers although it could also buy them from open markets. Only in this way it is possible to generate competitive advantages over its competitors.

Uncertainty poses a transactional problem of a different nature. It is part of the decision environment in which transactions take place and refers in general to situations where the relevant decision contingencies cannot be spelled out *ex ante*. In a PSM context, a particular source of uncertainty is volume unpredictability resulting partly from volatility in the buyer's downstream market. This form of uncertainty creates an adaptation problem which in turn gives rise to transaction costs in connection with modifying contracts to the new circumstances (Christopher *et al.*, 2000; Heide *et al.*, 1995). The more specific investments are required and the higher the uncertainty is, the more appropriate are coordination, collaboration or formal integration compared to open market relationships. Transaction cost theory shows that supplier-buyer relationships based on ongoing relationships can make sense from an economic point of view under the conditions of high uncertainty and transaction-specific investments.

In summary, transaction cost theory implicates that supplier-buyer relationships should be coordinated in a selective fashion, based on careful attention to the existing influential factors and implications for transaction costs and the possibilities to generate competitive advantages. It proposes that a buyer and a supplier will develop an ongoing relationship if the costs associated with maintaining this specific relationship are less than the costs of buying in an open market or a formal integration (Monczka *et al.*, 1995). Furthermore, transaction cost theory contributes to the development of total cost of ownership (TCO) concepts through highlighting the relevance of non-production costs. PSM managers might especially integrate the specificity investment insights into their supplier selection and development activities. Based on transaction cost theory, PSM managers might adapt their supplier relationships in order to achieve optimal efficiency and effectiveness through power or cooperation. .

2.2.3 The systems theory-based new St. Gallen management model as a framework for PSM

The generation of competitive advantages is a complex and multi-faceted challenge because business relationships are increasingly dynamic and interdependent (Rüegg-Stürm, 2005). This dynamic environment causes that experience and apparent regularities do not often support successful management practice anymore. Similarly, management concepts without strong theory background provide only little value to research and practice. In this doctoral thesis, the new St. Gallen management model provides a framework that supports the systematic break-down of companies' strategic priorities into the strategy-based selection of suppliers.

Companies can no longer be seen as trivial systems. A systemic and networked concept is necessary to master complex management challenges (Ulrich, 1985). Systems theory (e.g. Bertalanaffy, 1995) provides a possible approach to cover with complex environments (Jahns, 1999). Systems theory is focused on a holistic and integrated problem view to cover with complexity and avoid a limited perspective only on parts of a problem. The starting point of systems theory is the assumption that social systems are constructed through the preservation of differences to their complex environments (Luhmann, 1996). Systems are defined as structured relational fabrics trying to select specific possibilities from the environment and avoid others (Kühl, 1995). They are always in contact with their complex environments and are therefore forced to focus on specific parts of them (Schreyögg *et al.*, 1985). Otherwise, systems could not reduce the complexity of their environments. This compulsion to select a specific system-environment-difference through the system itself is one of the most significant differences between new systems theory and classical systems theory approaches (Jahns, 1999).

New systems theory defines companies as complex systems consisting of numerous interrelated elements. Companies have a specific institutional character and are more than the sum of their elements (Schreyögg, 2000). The identity of a system is mainly built on the differences to its environment and arises from the limitations of the system compared to its environment. Systems are forced to defend their chosen part of their environments against other possibilities in order to keep their identity (Kühl, 1995). This permanent selection of parts of the environment creates uncertainty which can cause a separation of the system from significant developments in the environment. This constitutes a risk for the system that needs to be compensated (Jahns, 1999).

Another element of systems theory is the management of complexity through the build-up of subsystems within the system (Luhmann, 1996). Systems can cover with complexity in their environments only if they are sufficiently complex as well. The subsystems provide the required potential to manage with the complexity of the environment because each subsystem covers a specific part of the external complexity. These subsystems provide a higher complexity management capacity for the system because external turbulences are often already managed in one of the subsystems therefore achieving more stability in the system (Schreyögg, 2000). These subsystems manage themselves; i.e. subsystems are like viable cells which are required to determine their function for the system on their own (Jahns, 1999).

Based on systems theory (e.g. Luhmann, 1996) and other systemic-constructionist approaches to management (e.g. Giddens, 1984; Ulrich *et al.*, 1984), scholars of the University of St. Gallen have further developed their management model (Rüegg-Stürm, 2005). The new St. Gallen management model conceptualizes companies as complex systems.

The new St. Gallen management model consists of six central categories relating to different management perspectives (figure 9):

(a) Environmental spheres (society, nature, technology, economy)

(b) Stakeholders (e.g. employees, customers, investors, government)

(c) Interaction issues (concerns, interests, norms, values, resources)

(d) Structuring forces (strategy, structure, culture)

(e) Processes (management processes, business processes, support processes)

(f) Modes of development (optimization, renewal)

The following sections will focus on two categories: structuring forces (d) and processes (e). These two perspectives on management are the most relevant for the research approach in this doctoral thesis and provide valuable insight into the interaction between structures and processes.

The structuring forces of a company are strategy, structure and culture. Although especially the strategy element is important for this doctoral thesis all three structuring forces are briefly discussed. Firstly, strategy is concerned with providing orientation knowledge (what?) which ensures that companies' efforts are directed towards the objectives defined as success. Secondly, structure gives a company the necessary coherence and provides coordination for the different value creation processes. It is about how things are done in a company. Finally, culture provides a common sense of purpose (why/what for?) which goes beyond the strategic and structural arrangements of the company. It provides an anchor when people are required to understand and react upon ambiguous or even contradictory events (Rüegg-Stürm, 2005).

Figure 9: The new St. Gallen management model

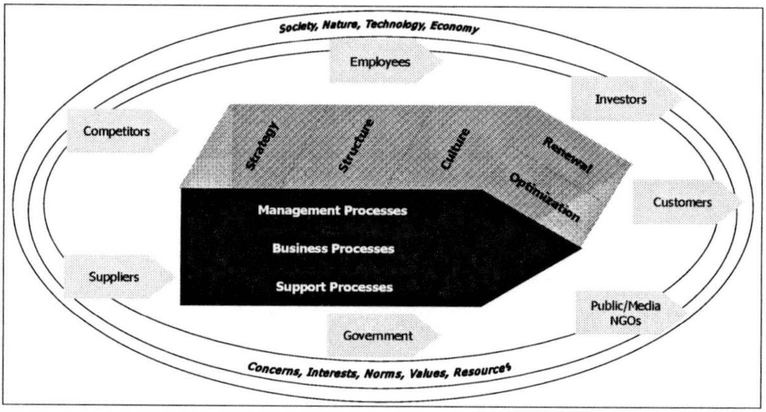

(Source: Adapted from Rüegg-Stürm, 2005)

Following the research objectives of this doctoral thesis, the structuring force strategy is of high interest. According to *Rüegg-Stürm* (2005), strategy is concerned with systematically examining the foundations for long-term success of a company in terms of competitive advantages. Among others, strategy has to provide detailed information about the company's market positioning model, its value proposition model and the required capabilities leading to strategic orientation knowledge within the company. This understanding of strategy is also applied in this doctoral thesis.

Just as important for the research concept of this doctoral thesis are the basic processes of a company. Processes can be understood as systems of activities which are accomplished in a more or less standard sequence and create an output. The new St. Gallen management model classifies three main processes (Rüegg-Stürm, 2005):

(1) Management processes

(2) Business processes

(3) Support processes

Ad (1) Management processes

Management processes cover all fundamental management tasks concerned with designing, controlling and developing a purpose-oriented socio-technical system (Ulrich, 1984). They are further separated into normative orientation processes, strategic development processes and operative management processes. Normative orientation processes focus on dealing with the various stakeholders devising a code of conduct in cases of conflicting interests. Operative management processes cover mainly the management of the day-to-day business processes. Most interesting for this doctoral thesis are the strategic development processes which consist of all activities leading to the development and implementation of a strategy.

Ad (2) Business processes

From a company perspective, business processes comprise customer processes, supply chain management processes and innovation processes. Firstly, customer processes consist of three sub processes customer acquisition, customer retention and brand management. Secondly, supply chain management processes include all activities necessary to perform a service or manufacture a product. Thirdly, innovation processes comprise all sub processes which contribute to systematic product and service innovations.

Ad (3) Support processes

Support processes include all sub processes which make the infrastructure available and provide the necessary internal services in order to efficiently and effectively accomplish business processes. Examples are infrastructure care or law services.

Due to the fact that the new St. Gallen management model is strongly based on systems theory, this general management model can also be applied to subsystems of a company such as PSM (Jahns, 2005b). It provides managers with different perspectives on companies but also on PSM as a subsystem similar to *Jahns'* (2005b) supply management navigator. Especially the management perspectives structuring forces and processes provide valuable input for PSM research and practice. In particular, the interaction between the structuring force strategy and the strategic development process as part of the management processes is of high interest for this doctoral thesis. The strategic development of PSM largely follows specific procedural patterns which can be traced back to PSM's already existing strategy, structure and culture. The result of this strategic development is a reworked PSM strategy with an appropriate structure and culture. Therefore, the result of the strategic development process becomes itself a part of the structuring forces. Accordingly, there exists a circular interaction between structuring forces and business processes on PSM level (Rüegg-Stürm, 2005). Furthermore, the process concept can be applied to PSM in particular while distinguishing management processes, business processes and support processes within the PSM system. In connection with the structuring force strategy it becomes apparent that an overall PSM strategy needs to be split up into strategic initiatives (Rüegg-Stürm, 2005) for all three value creation processes.

2.3 PSM's integration into strategic management

The literature about strategy and PSM can be divided into three distinctive areas (Ellram *et al.*, 1994): PSM's strategic relevance, PSM's possibilities to achieve competitive advantages and PSM strategies. In the following, these aspects are discussed in order to gain a better understanding of PSM's possible integration into strategic management and how it can create competitive advantages through a strategy-based selection of strategic suppliers.

As some authors (e.g. Dierickx *et al.*, 1989; Ramsay, 2001a) claim that PSM has no significant strategic role to play and its activities are mostly operational in nature, a short look at the relevance of PSM from a strategic point of view seems appropriate before discussing the central aspects of strategic PSM. The critics of a strategic role for PSM argue that profit contribution alone does not imply strategic importance as every function in a firm can claim to contribute to profits through cost-reducing activities (Ramsay, 2001a). They use a resource-based view to show that PSM cannot provide the company with sustainable[13] competitive advantages arguing that they derive from within a company and not from activities performed between companies and their suppliers.

[13] Sustainable and valuable resources are generally defined by the following criteria: rare, non-substitutable, and imperfectly imitable (Barney, 1991, pp. 105-108).

According to *Gälweiler*, the most decisive criterion to be perceived as strategic is not profits but the development or maintenance of success potentials (Gälweiler, 1987). Therefore, if PSM can be shown to fulfill this criterion we can claim its strategic character. The argument of PSM's operational character does not preclude a function from being finally strategic because there also are such things as an operations or production strategy which is part of companies' wider strategy system. *Carter* and *Narasimhan* (1996) prove with their empirical study that PSM is just as important as marketing, finance, accounting, or other operational issues to a company's competitive success. Most proponents (Goebel *et al.*, 2003; Mol, 2003; Spekman, 1985) of a strategic role of PSM base their arguments on a resource-based view but also apply other perspectives such as *Porter's* (1985) five forces model or the relational view (Dyer *et al.*, 1998b) which will shortly be discussed in the following.

As some critics use the resource-based view to deny PSM's strategic character some proponents use the same theory to show its strategic relevance. In short, the supporters of a strategic role claim that the resource-based view has not even in its original concept effectively ruled out achieving sustainable competitive advantages through outside partners such as suppliers. Although *Barney* (1991) explains that companies cannot purchase sustainable competitive advantages in open markets he also states that such advantages must be found in the rare, imperfectly imitable, and non-substitutable resources *controlled* by a company. He does not demand to *own* the resources but only to *control* them. This implies that buying firms can achieve sustainable competitive advantages if it is able to use a suppliers' valuable resources and *control* them by supply guarantees, long-term contracts or mutual commitment (Mol, 2003; Ramsay, 2001b).

At least three of *Porter's* (1985) five forces concept involves companies' suppliers and therefore PSM. Bargaining power appears to be the most obvious force related to PSM and competitive advantages. Buying firms may be able to find ways to accumulate more bargaining power towards suppliers by bundling the requirements of different business units. A second force related to suppliers is the threat of new entrants which can happen through suppliers' downstream investment. Managing supplier relationships and preventing downstream competition thus can be a strategically important activity for the company. Finally, substitutes as the third force can be developed by suppliers using specific knowledge maybe even derived from the buying firm. From this perspective, it doesn't seem possible to conclude that PSM has no strategic role in generating competitive advantages (Ellram *et al.*, 1994; Mol, 2003). Another prominent argument in favor of PSM's strategic role is based on the relational view. One of its core messages is that relations between companies and their suppliers can lead to superior performance and competitive advantages due to idiosyncratic relations. The combination of resources in unique ways and the development of trust may realize an advantage over competing firms which are unable to do so (Dyer *et al.*, 1998b; Johnston *et al.*, 2004). As the relational view assigns a strategic role to supplier management PSM needs to be perceived as strategic as well (Mol, 2003). Having shown that PSM is worth being analyzed from a strategic perspective, a closer look at this issue seems necessary discussing the role of PSM in strategic management.

In general, strategic management is divided into two basic perspectives. Firstly, strategic management can be analyzed from a *process* (e.g. Mintzberg *et al.*, 1985) perspective focusing on the development process of all kinds of strategies. This research stream tries to explain how strategies are developed over time and which factors influence the strategy development process. Secondly, strategic management questions may also aim at the formulation of strategy *content* (e.g. Hitt et al., 1998). Content research in strategic management focuses on the answer which possible strategic priorities offer companies optimal results under varying environmental conditions (Müller-Stewens *et al.*, 2003). PSM can be analyzed accordingly. Therefore, the first chapter is concerned with the question of how to develop a PSM strategy while the second chapter tries to answer which content belongs to a PSM strategy. The first chapter is important to further understand the break-down process of strategic priorities from top-level to bottom level. The second chapter supports the understanding how PSM strategies are structured and how a strategy-based supplier selection can be supported. *Carr* and *Smeltzer* (1997) as well as other authors (e.g. Large, 2000; Zsidisin *et al.*, 2003) distinguish the topics accordingly calling strategic purchasing or strategic procurement all issues which refer to the planning process PSM follows as part of the strategic management process (Carr *et al.*, 1999b; Pearson *et al.*, 1990). Therefore, strategic PSM as a comprehensive term refers to the strategic process analyzing the environment and internal structures, setting goals, developing alternatives, evaluating alternatives, and implementing as well as controlling the chosen alternatives in PSM. Purchasing strategy, supply strategy or PSM strategy as a comprehensive term relates to the specific objectives PSM may take from a content perspective. This includes all goals and activities concerning suppliers, commodities, supply markets, internal customers, PSM staff or top management. Strategic PSM process aspects and PSM strategy content aspects will be analyzed and discussed in the following two chapters. This analysis serves as the basic concept for the break-down process from companies' top-level priorities into supplier selection activities. In the last chapter of 2.3, supplier capabilities will be discussed as a possible evaluation approach for a strategy-based supplier selection and facilitator for the break-down process.

2.3.1 PSM's former and new role in companies' strategic management process

As described in the practice gap analysis in chapter 1.2, PSM has still a clerical role in many companies (Jahns, 2005b; Moody, 2001). Those clerical PSM organizations can be characterized as an operations support function which has a short-term focus and reacts on quality and cost problems (Monczka *et al.*, 2002). Many of recent research activities within the PSM field have been concerned with the examination of purchasing's clerical role and the development of supply management's strategic role (Harland *et al.*, 1999). There are several models which describe the evolution from clerical purchasing to strategic supply management (Bhote, 1989; Hines *et al.*, 2000; Monczka *et al.*, 2002). As a first example, the evolution model from *Dobler* and *Burt* (1996) is presented in figure 10.

Figure 10: Evolution steps from reactive to strategic procurement

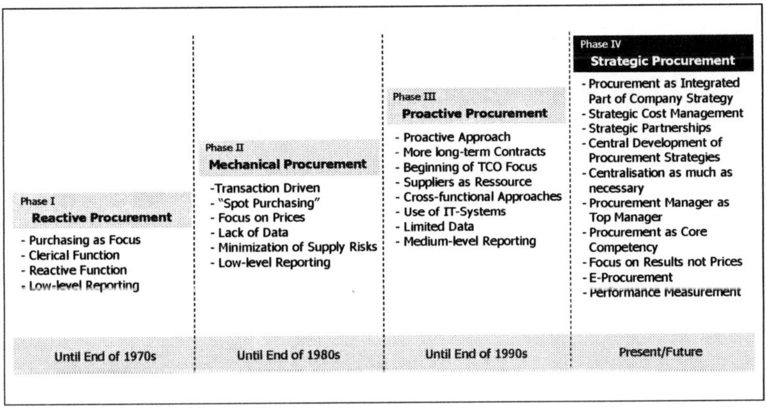

(Source: Adapted from Dobler *et al.*, 1996).

According to their model, PSM used to be characterized as a reactive function with low-level reporting until the 1970s. Receiving the orders from other functions, PSM was focused on transaction activities concentrating on achieving the lowest possible prices. Other dimensions such as quality or innovation were not in PSM's primary decision responsibility. It took PSM almost 30 years to move towards a more strategic function accepted as a peer to marketing, production of finance and described as strategic procurement. In this last stage, procurement is described as an integrated part of corporate strategy with centralized procurement strategy development. Only within such a context, an adequate integration of PSM into companies' strategic management processes is possible.

Another evolution model is the four stage concept developed by *Reck* and *Long* (1988). They analyzed PSM's contribution to companies' competitive strategies in 15 case studies. Similar to *Scheuing* (1998), they stress PSM's evolution from a pure price focus on materials to a value orientation for the overall business success (figure 11). Once again, only the last development stage describes PSM as a function accepted by its peers. The change from a facilitator (stage III) to a peer (stage IV) points out to the important change in a company's mindset that PSM can create competitive advantages on its own and is not primarily as supporter of other functions.

Figure 11: Four stages of purchasing development

Stage I - Passive	In the passive stage, purchasing normally begins as a reactor to requests from the other departments. Many of purchasing's legitimate activities are handled by other functions outside of purchasing.
Stage II - Independent	In the independent stage, purchasing departments spend considerable time attempting to professionalize in the purchasing function by introducing such things as computerized information systems, formalized supplier programs, and communication links with the technical functions.
Stage III - Supportive	In the supportive stage, purchasing departments are viewed by top management as essential business functions. Purchasing is expected to support and strengthen the firm's competitive advantage by providing timely information to all departments in the firm about potential changes in the price and availability of materials, which may impact the firm's strategic goals.
Stage IV - Integrative	In the integrative stage, the firm's competitive success rests significantly on the capabilities of the purchasing department's personnel. Purchasing's role within the firm changes from facilitator to functional peer. This development process must be implemented and guided by management over a period of time

(Source: Adapted from Reck *et al.*, 1988).

In addition, *Freeman* and *Cavinato* (1990) developed an influential evolution model for PSM (figure 12). It is based on four stages similar to the seminal model of *Gluck*, *Kaufman* and *Walleck* (1980). *Freeman's* and *Cavinato's* (1990) main research interest was to evaluate whether PSM's development stage varies correspondingly with companies' overall development stage. One of their main research results is that companies' stage of development closely correlates with the PSM's stage of development (Freeman *et al.*, 1990).

PSM organizations in phase I of their model (figure 12) are characterized as providing buying services within a company. PSM's main objective is the minimization of costs against the budgetary norms. The goods and services bought are generally limited to such things as office supplies or MRO items. PSM organizations in phase II place a strong emphasize on forecasting and include cost avoidance considerations. Quality issues and professional human resource management for the PSM organization emerge. Phase III in PSM development includes the responsibilities for inventory, transportation and outsourcing issues. It is focused on the support of the lines of business. PSM plans tend to be integrated with other functions' plans.

In phase IV, PSM integrates the broad concept of supply management and anything that involves material or service needs is included within the scope of PSM's responsibility (Freeman *et al.*, 1990). Large companies might incorporate all four phases in their PSM organizations. For instance, while a buying site that reports to a plant manager is characterized by phase I, the same company has a centrally managed group of PSM managers who coordinate the exchange between innovative suppliers and the R&D department (Freeman *et al.*, 1990).

Figure 12: Purchasing emphases throughout the range of strategic settings

Purchasing Attributes	**Stage I** Basic Financial Planning	**Stage II** Forecast Based Planning	**Stage III** Externally Oriented Planning	**Stage IV** Strategic Management
Concept of the Field	**Buying**	**Purchasing**	**Procurement**	**Supply**
Concept of "Strategy"	Better price on next buy	Maintain favorable price / cost variances	Support line of business	Entrepreneurial team member
Expectations	Minimization of costs	Cost minimization, reduction, avoidance Purchase for quality	Contributions through value analysis, value engineering	Involved in product development and line of business management, line of business results
Management Approach	Reactive	Reactive but plan for future	Fit department in with plans of rest of firm	Positive pro-active
Major Activities	Process requisitions into purchase orders and contracts	Management of the buying function Make process efficient	Fit buying cycle to the line of business product cycle	Manage commercial relationships for the firm, source for long term
Range of Products	MRO items, office goods	Raw materials, MRO items, office goods	Capital goods, raw materials MRO items, office goods, outsourcing management	Suggest source firms to purchase, suggest product changes in line with market opportunities and future constraints
Budgetary Approach	Cost center	Cost center, planning for future	Supply chain management, shape future of department for line of business or SBU	True supply management, partner in change
Management Style	Clerical / reactive	Managerial, forecasting	Managerial, planning	Team member
Key Personal Skills	Task oriented	Some management	Managerial, strong interpersonal and analytical	Purchasing decisions are business decisions
Concerns	Conformance to norms, Process problems	Basic managerial issues, concern with power regard-ing scope, back door buying, headcount, centralization etc.	Supply chain management	Shape of function not important, results and output are the keys

(Source: Adapted from Freeman *et al.*, 1990).

Most evolution models assign a strategic character to PSM only to their final stages. In the last evolution stage PSM is integrated into a company's strategic management processes and is able to directly create competitive advantages (Paulraj *et al.*, 2005). For this purpose, PSM needs to be linked with the corporate and business unit level objectives. They provide the relevant guidelines for the PSM strategy on the functional level assuring that it is directly supporting company objectives and not primarily other functional priorities, for example, from production or R&D (Carr *et al.*, 1997).

The relationship between corporate level strategies on the one hand and the PSM strategy as a functional strategy on the other hand is not one-dimensional (Pagell *et al.*, 2002). Although not part of this doctoral thesis PSM may also have a major influence on the decisions made on corporate and business unit level. Some authors (e.g. Browning *et al.*, 1983; Freeman *et al.*, 1990) identified areas in which PSM can contribute to corporate strategies. For example, supply market trends are monitored and exploited in order to support the company's overall objectives. Additionally, providing an overview of materials, services and technologies available on the supply market as well as developing new supplier options can support the development of competitive advantages. As a result, PSM managers have to be heard and integrated in the core group for strategic development on corporate and business unit level in order to maximize short-term and long-term opportunities. By this way, companies can make decisions based not only on marketing's identification of customer needs or R&D's technology push, but also on PSM's ability to identify and source improved materials and services and consequently develop new business solutions (Evans *et al.*, 2001; Pearson *et al.*, 1990).

This need for integration is also supported by *Pearson* and *Gritzmacher* (1990) who highlight the importance of PSM's integration into strategic management in their studies. Their analysis of the content of corporate or business unit strategies within the strategic management literature failed to identify any significant contribution of PSM to the development of competitive strategies (Pearson *et al.*, 1990). As a consequence, they require the integration of PSM as an equal contributor along with marketing, finance or production into the main strategic management processes.

2.3.2 Supplier selection as part of PSM strategy

The content of a PSM strategy has changed over time due to the evolutionary development from purchasing to supply management as shown in the previous chapter. As already defined, the term PSM is used to cover all clerical and strategic aspects. By analogy with this comprehensive understanding, PSM strategy covers all relevant strategy content aspects. This is a broad definition of PSM strategy which especially requires the systematic integration of supplier selection in order to align supplier selection priorities with company requirements. In the following, it is shown how a PSM strategy can be organized content-wise and which PSM strategy elements are decisive for the integration of a strategy-based supplier selection.

Based on *Jahns'* (2005b) supply management navigator concept management activities, core activities and support activities in PSM are distinguished. This differentiation corresponds with the distinction of processes in the new St. Gallen management model described in chapter 2.2.3. Management activities cover topics such as organizational issues, risk management or human resource management in PSM. Support activities include mainly enablers such as e-procurement or value analysis tools. Core activities in PSM include all aspects concerning suppliers, supply markets and internal customers. Before developing a new PSM strategy model, some approaches from literature are presented and analyzed in order to ensure the integration of all relevant aspects.

A comprehensive procurement strategy approach has been developed by *Arnold* (e.g. 2002) who proposes six different procurement strategy elements:

(1) Lieferantenstrategie[14]

Supplier strategy defines the number of suppliers for a specific component. The basic possibilities are sole, single, dual, and multiple sourcing. Sole sourcing happens if a supplier has a monopolistic role in a specific supply market (Owens *et al.*, 1994). Single sourcing exists if a buying firm voluntarily chooses only one supplier (Arnold, 2002; Schorr, 1998). Dual sourcing includes two suppliers for a specific commodity. Finally, multiple sourcing is defined as sourcing a specific commodity from more than two suppliers.

(2) Beschaffungsobjektstrategie[15]

The procurement object strategy is about what suppliers deliver. The basic alternatives are raw material, units, modules and systems. Classically, companies buy units from suppliers but as the complexity of production processes increases, companies source more modules and systems from their first tier suppliers. In contrast to modules, systems sourcing usually integrates some aspects of R&D activities (Gadde *et al.*, 2002; Wolters, 2002).

(3) Beschaffungszeitsstrategie[16]

The procurement time strategy aims at optimizing the stock situation. Possible solutions are stock sourcing, demand tailored sourcing and just-in-time sourcing (Stölzle *et al.*, 2002). Stock sourcing highlights the relevance of supply reliability and security in case of supply interruptions or other problems. Buying firms often require a specific stock level at the place of production or assembly. Demand tailored sourcing tries to balance cost and security aspects while just-in-time sourcing explicitly tries to integrate the different members of the supply chain in order to optimize stocks not only for the focal company (Arnold, 2002).

---

14 Supplier strategy

15 Procurement object strategy

16 Procurement time strategy

(4) Beschaffungssubjektstrategie[17]

The procurement subject strategy answers the question in which organizational form the procurement managers act. The proposed two basic alternatives are individual and cooperative sourcing.

(5) Beschaffungstechnologiestrategie[18]

The procurement technology strategy focuses on the information and communication technologies such as procurement cards or e-auction tools used in the sourcing process (Arnold, 2002).

(6) Beschaffungsarealstrategie[19]

The procurement area strategy defines the geographical area where the buying firm is actively searching, selecting, evaluating and integrating suppliers. The most important dimensions are local, domestic, regional and global sourcing.

While *Arnold's* (2002) approach reveals some interesting aspects there seem to exist some inconsistencies. For instance, if the sourcing subject belongs to the main part of a procurement strategy then the procurement structure should play a major role as well, including organization and process aspects. Compared to *Jahns'* (2005b) supply management navigator topics from all three levels are included. While supplier strategy, sourcing object strategy and sourcing area strategy belong to the core activities and the sourcing subject strategy can be ascribed to the management level, sourcing technology strategy and sourcing time strategy belong to the support level. Although elements from all three levels are represented no clear cut-off criterion or system is provided.

Koppelmann (2002b) developed another strategy concept following an objective-oriented approach. He defines the elements of a procurement strategy as follows:

(1) Produktstrategien[20]

Product strategies include a collection of possibilities how to manage the sourced products of a company. Some of the dimensions mentioned include the cooperation with suppliers, the complexity of the sourced object such as unit, module or system, and the degree of standardization.

---

17 Procurement subject strategy
18 Procurement technology strategy
19 Procurement area strategy
20 Product strategies

(2) Bezugsstrategien[21]

Acquisition strategies focus on supplier-related aspects such a the geographical scope ranging from local to global, the supply market mix and degree of market concentration, the number of suppliers for a specific product or service, the way of supply, and the determination of the sourcing subject towards suppliers.

(3) Kommunikationsstrategien[22]

Communication strategies consist of an information acceleration strategy which aims at fostering the information exchange with suppliers and the development of a weak signals strategy focusing on the early detection of problems with suppliers.

(4) Servicestrategien[23]

Service strategies include performance delegation, disposal delegation and control intensification. All these aspects are closely related to the outsourcing of functions whereas disposal delegation is a special case of performance delegation. Control intensification is concerned with the control problems in outsourcing relationships.

(5) Preisstrategien[24]

Price strategies focus on the optimal price determination of products and services. Minimal price strategies aim at reducing the price paid for a product to the lowest possible level while fair price strategies try to balance a low price level and the motivation of suppliers to further invest in their competitiveness for a specific product or service (Schorr, 1998). Also possible is the market price strategy which is especially applicable for commodities traded on open markets.

Koppelmann's (2002b) approach includes many different aspects from outsourcing decisions to negotiation strategies but focuses on an objective-oriented view. Management activities as proposed from *Jahns* (2005b) are not expressively included limiting the strategy content to core activities compared to the supply management navigator concept.

Leenders, Fearon, Flynn and *Johnson* (2002) also developed a supply strategy concept. They propose that any supply strategy should include a determination of 'what', 'quality', 'how much', 'who', 'when', 'what price', 'where', 'how' and 'why'. 'What' implies the fundamental question of make or buy as well as the issue of whether to buy standard items

21 Acquisition strategies
22 Communication strategies
23 Service strategies
24 Price strategies

and materials readily available in the market or custom-specified goods. 'Quality' focuses mainly on the trade-off problem of quality versus cost. 'How much' as another major component of their supply strategy concept deals with the question of how much is to be acquired in total and per delivery. The possibilities range from large stocks to just-in-sequence delivery. 'Who' focuses on organizational issues such as the degree of supply organization centralization, purchasing staff quality or top management involvement. The 'when' question is closely related to the one of 'how much' and brings up issues such as forward buying and inventory policy. 'What price' is concerned with concepts such as cost-based versus market-based approaches as well as premium prices versus standard prices. *Leenders, Fearon, Flynn* and *Johnson* (2002) subsume under 'where' diversified concepts such as local versus global, large versus small, single versus multiple and supplier relations aspects. This accumulation of concepts under 'where' seems not entirely understandable. The component 'how' also represents such an accumulation including systems and procedures as well as negotiations, material requirement planning or ethics. Finally, 'why' is concerned with the question why a specific supply strategy should be pursued. This issue provides a link to objectives at business unit or corporate level. Compared to *Arnold's* (2002) approach, many similarities between the two concepts can be identified. For example, 'where', 'who' or 'when' are closely related to the sourcing area strategy, sourcing subject strategy and the sourcing timing strategy. Compared to *Jahns'* supply management navigator model, all proposed strategy elements except 'why' might be attributed to the core or support module.

Taking another perspective, *Spekman* (1985) proposed three levels for strategic procurement planning:

(1) Competition-related strategy

Competition-related strategies focus on PSM managers' market power which allows them to lever up procurement and to improve companies' competitive advantages. Key aspects to achieve market power are supply market intelligence, supply base management, make-or-buy analyses and minimization of supplier dependence (Virolainen, 1998).

(2) System-related strategy

System-related strategies build information links between the buying firm and its more immediate external environment. These mid-level strategies focus on topics such as supplier selection, inventory management, value analysis or multinational sourcing.

(3) Performance-related strategy

Performance-related strategies focus on managing purchasing resources, controlling expenses and serving internal customers' needs. Their main purpose is the achievement of purchasing efficiency.

Spekman's (1985) concept highlights the importance of an integrated approach to PSM strategy. His distinction is based on a similar logic like *Jahns'* (2005b) approach. His performance-related strategy shows the importance of a focus on management topics in PSM. Furthermore, he links PSM to the generation of competitive advantages but separates PSM core processes such as supplier selection, commodity management or multinational sourcing.

The analysis of the described models presents a wide range of possible aspects concerning the content of a PSM strategy. Although the models present well specified content of PSM strategy each of them focus quite narrowly on some issues. According to *Virolainen* (1998), an integrated PSM strategy should include issues such as value chain positioning, organization structuring, make-or-buy decisions, choice of supplier strategies and links to corporate level strategy or other functional strategies. Therefore, an approach is necessary categorizing the different PSM strategy elements in a systematic way. *Spekman's* (1985) approach might serve as a starting point for a PSM strategy categorization.

In this doctoral thesis, the basic distinction of *Jahns'* (2005b) supply management navigator concept is applied to define the strategic components in PSM. Similar to the new St. Gallen management model presented in chapter 2.2.3 (Rüegg-Stürm, 2005), he distinguishes a management, core and support level. In analogy with strategy development in marketing (Backhaus, 1999; Becker, 1998) a PSM strategy is conceptualized as a combination of different sub strategy elements (Arnold, 2002).

On the top level, a PSM strategy includes all relevant aspects to achieve the goals by applying all necessary activities from clerical purchasing to strategic supply management. On a second level, three basic sub strategies are defined based on the management, core and support process logic (Jahns, 2005b; Rüegg-Stürm, 2005). This implies that a PSM strategy consists of three main elements:

(1) Supply strategy (management level)

(2) Sourcing strategy (core level)

(3) Support strategy (support level)

Ad (1) Supply strategy

According to *Harland, Lamming* and *Cousins* (1999), the concept of supply strategy forms a holistic, strategic management perspective of operations, stretching across inter-organizational boundaries. Their understanding of supply is therefore broadly defined. They define the concept of supply as a holistic approach to manage activities within collaborative inter-organization networks, allowing the use of rational strategies for the satisfaction of customer demand through innovations and network structures in a global environment (Harland *et al.*, 1999). Although *Harland, Lamming* and *Cousins* consider supply as a comprehensive term, the focus of the supply strategy lies on the strategic management activities. *Kauffman* (2002) adds that the determination of a supply strategy has been presented in several ways in the literature. A possible approach is first to identify corporate strategy in order to ensure that supply strategy supports it and then to select the most appropriate strategy components. Another approach is to segment the supply market using tools such as spend analyses and subsequently define the optimal strategies for each segment. *Jahns* (2005b) has developed a systems theory based supply management concept which identifies supply strategy as a specific module of the supply management navigator. Similar to *Harland, Lamming* and *Cousins'* (1999) approach, his supply strategy concept focuses on the relationship between PSM's strategic priorities and companies' strategic priorities fostering PSM's effectiveness (Jahns, 2005b). The supply strategy module represents best the relevance of general management topics in PSM.

Following the presented approaches, the applied supply strategy definition in this doctoral thesis contains all elements that require the application of general management principles to achieve PSM's strategic priorities. This includes for example supply controlling and risk management, supply organization and process management and especially PSM's integration into companies' strategic management processes. The supply strategy has to ensure that PSM's staff can work in an optimally organized environment fostering an effective and efficient implementation of the sourcing strategy.

Ad (2) Sourcing strategy

According to *de Quervain* and *Wagner* (2003), sourcing strategies are strategies for a specific part, service or module. According to *Kauffman* (2002), literature still lacks the identification of what, at the core, are the basic strategic elements which are applied in any definition or aspect of purchasing and supply management. He proposes the elements product, cost, relationship and access as the four key elements (Kauffman, 2002). Sourcing may also be defined as a set of processes that involves analyzing spending patterns and supply markets to find the optimal suppliers (Mitchell, 2002). In this approach, three key elements can be identified: analysis of major commodity groups, supply markets and suppliers. *Van Weele* (2002) suggests a similar approach relating the sourcing strategy concept mainly to supplier management also including supply market questions and commodity group considerations (van Weele, 2002).

In this doctoral thesis, the sourcing strategy approach is based on *Jahns'* (2005b) core module of the supply management navigator. The core activities of PSM are defined as supplier management, supply market management and internal customer management. These elements can also be found in the approaches from *Arnold* (1997) and other researchers (e.g. van Weele, 2002). They form the basic activities in PSM. In contrast to *Koppelmann's* (2002b) approach, the product view included in the internal customer management is not superior to the other elements. All three aspects have to be equally managed by PSM managers depending on the specific business context and companies' strategic priorities (figure 13).

A one-size-fits-all sourcing strategy makes little sense across a company's commodity portfolio. Suppliers might be plentiful and eager to compete in some categories while for other categories, supply options might be limited. Some commodities requested from PSM's internal customers may have a larger impact on business in terms of total expenditures or on companies' strategic position while the impact of other categories might be relatively small (A.T. Kearney Inc., 1998).

Figure 13: Sourcing strategy elements

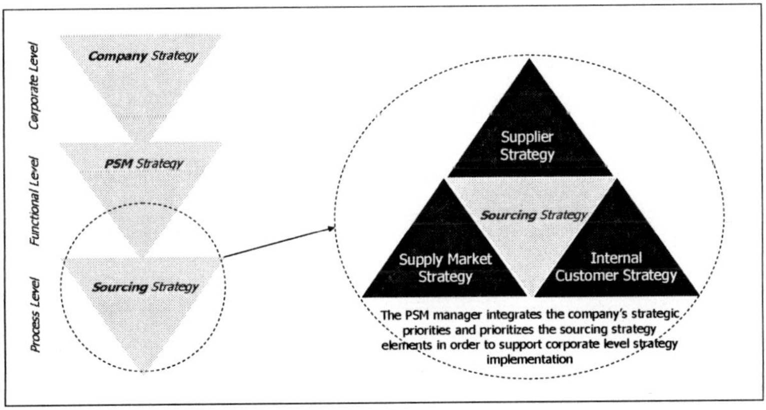

In many cases, the commodities requested by PSM's internal customers will drive the sourcing strategy while the supplier perspective or supply market perspective will be of secondary importance. In some cases, products or services might only be sourced from a specific supplier giving the relationship with this supplier the primary role in the sourcing strategy. Finally, factor cost advantages might lead to the decision to source a specific spend volume from a single supply market such as China or Czech Republic. Following this decision, internal customer management and supplier management are secondary to the supply market decision. Consequently, PSM managers have to integrate their companies' strategic priorities into the sourcing strategy and to decide about the relevance and detailed implementation of all three sourcing elements.

In order to gain a better understanding of the sourcing strategy concept its three major strategy elements are shortly described in the following:

(A) Supplier strategy

(B) Supply market strategy

(C) Internal customer strategy

Similar to the overall applied systems theory based PSM strategy concept (Jahns, 1999), the three elements of the sourcing strategy have to be seen as sub elements of the sourcing strategy.

Ad (A) Supplier strategy

Supplier management includes all activities necessary to develop and implement the sourcing objectives with the overall supply base and individual suppliers (Wagner, 2001). Key strategy elements are the number of suppliers and the kind of relationship such as arm's length transactions or collaborative partnerships (Ellram *et al.*, 1994). Furthermore, the number of suppliers for a specific product or service in the form of sole sourcing, single sourcing, dual sourcing or multiple sourcing is important. Supplier strategy defines the relevance of the applied selection criteria for strategic suppliers.

Ad (B) Supply market strategy

Supply market strategy includes all elements necessary to develop and implement the sourcing objectives in terms of supply market access and transparency. This includes not only the leverage of low cost country potentials to reduce costs of materials and services but also other objectives such as to gain exposure to worldwide product and process technology innovations, increase the number of available suppliers, satisfy counter trade and local content requirements or enter new customer markets (Monczka *et al.*, 2002; Zollenkop, 2003). Traditional research questions concern especially the geographical scope of sourcing activities such as local sourcing, domestic sourcing, regional sourcing or international sourcing. The term global sourcing is defined by many researchers (Trent *et al.*, 2005) as an independent sourcing strategy including more than geographical aspects. Global sourcing requires the integration of requirements in order to identify common purchases, processes, technologies and suppliers which can be coordinated (Bozarth *et al.*, 1998a). Based on the strategic concept of this doctoral thesis, global sourcing is not a comprehensive sub strategy for the sourcing strategy although it integrates many relevant aspects. However, the proposed strategic character (Arnold, 2002; Dobler *et al.*, 1996) of global sourcing compared to terms such as international purchasing draws the attention of PSM away from cost reduction objectives to such as innovation development or market entry considerations (Kotabe, 2002). Closely related to the geographical sourcing aspect are logistics requirements such as stock delivery, just-in-time delivery or just-in-sequence delivery.

Ad (C) Internal customer strategy

Internal customer management includes all activities necessary for developing and implementing the sourcing objectives towards PSM's internal customers. Relevant activities are standardization, bundling and specification management. Standardization tries to find or increase similarities between products and services which need to be sourced in order to improve PSM performance such as costs reductions or process improvements (Colsman, 2000; de Quervain *et al.*, 2003). Bundling efforts try to increase the sourcing volume for specific commodities across different business units or companies in order to improve the negotiation power towards suppliers. Specification management focuses on the examination of the real needs from internal customers towards required materials and services based on companies' overall strategic objectives. Internal customer management is also concerned with the determination how the needed material is sourced from a complexity level perspective. Standard forms in literature are raw materials, components, modules and systems (Gadde *et al.*, 2002).

Ad (3) Support strategy

The support strategy includes the configuration and application of all enablers as proposed in *Jahns'* (2005b) support modules. Its elements support primarily the development and implementation of the sourcing strategy and secondary the supply strategy. Examples are the intelligent integration of e-procurement applications and analysis tools such as supply portfolio management software or contract management databases.

These three strategy sub elements supply strategy, sourcing strategy and support strategy build the PSM strategy. This doctoral thesis focuses on the supplier strategy element as part of the sourcing strategy, particularly on the selection of strategic suppliers. As every other PSM activity, the selection of suppliers needs to support companies' strategic priorities and has to be aligned with the content of the other two sub elements of the sourcing strategy. This requires a break-down of companies' strategic priorities into PSM specific activities. In the following, a basic approach to a capability-based supplier selection is developed to support the required break-down process of strategic priorities.

2.3.3 Capabilities connecting supply and customer markets

The proposed break-down process from companies' strategic priorities into specific supplier selection criteria requires a transformation without any interface disruptions. Therefore, a common element is necessary which connects companies' strategic positioning towards customers and competitors with their supply markets.

While marketing or sales managers can talk about their "issues" in the same way as the company presents itself on the customer market or positions itself towards competitors PSM managers need to take a completely different perspective. If their company, for example, sells IT hardware and IT consulting services following an innovation leadership strategy this can not be directly translated into PSM's daily communication. The world PSM managers are mostly engaged with does not consist of innovative IT hardware and state-of-the-art IT consulting services but it consists of all the things that are necessary for the IT manufacturing and service providing processes. This different world therefore means a different language. Therefore, a challenge of today's PSM managers is to do the splits between their companies' product and service world and the world of their supply base where modules, systems, parts, raw materials or other sourcing units (Leverick *et al.*, 1998) are dominating the way of thinking (figure 14). Except in the retail industry, the world of purchasing and supply and the world of sales and marketing do not speak the same language often resulting in communication problems and a lack of mutual understanding. Consequently, the break-down process of strategic requirements from product-/market priorities and other strategies to the level of commodities or modules is a difficult task for PSM managers.

In contrast to marketing and sales managers, PSM managers must often bridge the gap between the 'product world' of many of their internal customers and the 'parts world' of their supply base in order to optimally source the required input. *Watts, Kim* and *Hahn* (1995) propose a capability-based approach which can build a bridge between the different perspectives. The offered products and services are the result of companies' and their suppliers' capabilities (Eilles *et al.*, 2003). Companies' market or competitive positions can also be traced back to capabilities. By this way, a consistent language and an uninterrupted information flow from companies' market offerings to the required input from the supply base can be fostered.

Figure 14: The 'World of Products' and the 'World of Parts'.

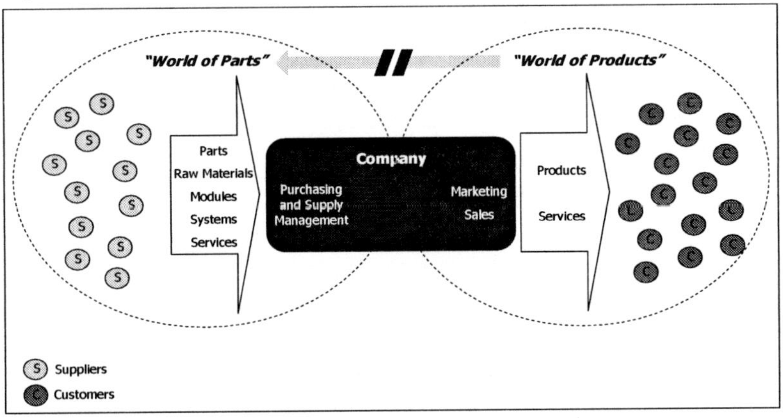

Following the same logic, *Talluri* and *Narasimhan* (2004) recommend a focus on supplier capabilities and practices when selecting strategic suppliers. The future selection of strategic suppliers will be much more focused on the suppliers' potential to reduce costs, to create new innovations or to reduce time-to-market. At Honda of America, many strategic suppliers are selected mainly based on the fact that their capabilities in technology or systems development can provide more input for Honda than it can currently imagine using in the immediate future. As competition is expected to become only fiercer, the potential of strategic suppliers to improve buying firms' competitive position will come into focus (Duffy, 1999).

For *Möller* and *Törronen* (2003), it is clear that the potential value of suppliers is highly related to their various capabilities. They and other researchers (e.g. Large, 2000) argue that it is a feasible alternative to examine the potential value of suppliers based on their current capabilities. In figure 15, *Möller* and *Törrönen* (2003) present an approach how capabilities are linked to the value production possibilities of suppliers.

Figure 15: Capability base and value production

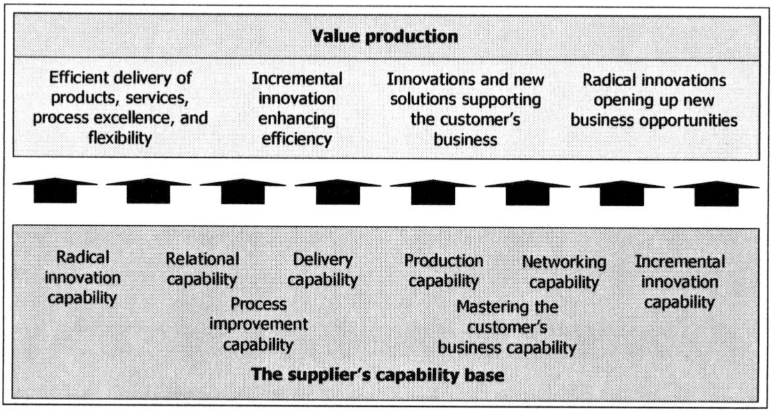

(Source: Adapted from Möller *et al.*, 2003).

In order to be managerially useful, their proposed capability approach needs to be operationalized. Clear indicators for each supplier capability which is considered essential in producing the required supplier value have to be derived. These indicators are often contextual, or dependent on the types of industry and technology in question. In the future, managers would benefit from having more context-specific capability indicators which should be identified by cross-functional teams with industry or technological experience. In summary, *Möller* and *Törronen* (2003) suggest that suppliers' value creation potential within buyer-supplier relationships can be examined through efficiency, effectiveness and network functions. These functions are interrelated but they are conceptually distinct. They conclude that buying firms should use suppliers' capability profiles as an overall indicator of how suitable a particular supplier is for a specific value creation project (Möller *et al.*, 2003). This conceptual approach will be applied in chapter 3 in order to develop a strategy-based supplier selection framework.

2.4 Interim results: Purchasing and supply management as a valuable contributor to the generation of competitive advantages

The analysis of PSM from a theory-based perspective provided more insights into its possibilities to contribute to the generation of competitive advantages. The analysis was based on the discussion of the most relevant terms and definitions. The term purchasing and supply management (PSM) was shown to be the most comprehensive in order to cover all necessary aspects discussed in literature. Furthermore, the strategic value of suppliers and the differences between resource, capabilities and competencies were discussed.

The analysis of PSM from four strategy theory perspectives and the new institution economics perspective provided valuable insights for research and practice about PSM's possibilities to generate competitive advantages and the appropriate selection of suppliers. Firstly, market-based view illustrated how PSM can add value through the build-up of entry barriers towards suppliers and the use of market power through creating supplier transparency and economies of scale. This classical PSM perspective also supported the understanding of when to use e-procurement tools. Next, resource-based view highlighted the importance of supply base analysis from a strategic perspective. The generation of Ricardian rents includes the knowledge of what is required from suppliers in the future and where to find these valuable supplier resources. This iterative relationship between PSM's possible contribution to business development and valuable supply market resources supports PSM's strategic character. Moreover, capability-based view focused on the aspects of generating temporary competitive advantages. Its main contribution was the focus on the relevance of innovations for the long-term success of a company. From a PSM point of view, the ability to identify and integrate innovative suppliers and create flexible buyer-supplier relationships bringing together complementary capabilities is of pivotal importance. Finally, relational view explained the value of idiosyncratic relationships and their requirements. It contributed to the value explanation of trust-based and long-term relationships as often discussed in current PSM literature.

Moreover, principal-agent theory provided useful insights into the question of how to optimize supplier management aspects such as the variety of selection criteria or complexity of contract terms. The situation of hidden characteristics showed the importance of reliable supplier capability indicators and performance measurement. The situation of hidden intentions exemplified the importance of trust and independence considerations in supplier selection. Finally, the concept of hidden actions explained the increased application of risk-rent sharing agreements in buyer-supplier relationships. Furthermore, transaction cost theory implicated that buyer-supplier relationships should be organized in a selective fashion, based on careful attention to the existing influential factors such a specificity of investments or transaction uncertainty. Based on transaction cost theory, PSM managers understand that buying non-strategic commodities from open markets is not always optimal.

The systems theory based new St. Gallen management model was analyzed as a potential framework for PSM. The focus of the analysis included the structuring forces and process perspectives strongly supporting a systems theory based PSM strategy concept. Following this theory-based understanding, PSM's integration into strategic management was further analyzed. Firstly, a process view was taken to discuss the interaction between PSM as a function and strategic management on corporate level. The main conclusion was that PSM needed to be more integrated into strategic management processes in order to generate competitive advantages. Next, a systems theory based PSM strategy content framework was developed to facilitate the break-down process from strategic priorities on corporate level into detailed activities on PSM level. Finally, it was shown that a capability-based measurement approach supports a strategy-based selection of strategic suppliers.

In summary, chapter 2 discussed the integration of PSM into strategic management. Although it was shown that PSM contributes to the generation of four rent forms (monopoly rent, Ricardian rent, Schumpeterian rent and relational rent) PSM is still not appropriately integrated into the strategic management processes in many companies. Consequently, PSM managers might primarily focus on their own contribution to company success through their supply base and secondary on the support of other functions. Focusing on supplier management, this primary focus on the generation of competitive advantages is only possible if the selection of strategic suppliers is aligned with the strategic priorities on corporate level. Therefore, PSM needs to develop a strategy-based supplier selection framework allowing the careful balance between strict alignment with the corporate strategy and the integration of other functional priorities from PSM's internal customers. The requirements and implications for such a strategy-based supplier selection are discussed on the following pages.

3 Strategy-based supplier selection

The most progressive companies pay close attention to their PSM function and attempt to improve the management of their supply base recognizing its contribution to strategy implementation and to overall corporate performance (Carr *et al.*, 1999a; Kannan *et al.*, 2002; Morash, 2001; Smith-David *et al.*, 1999; Vonderembse *et al.*, 1999). Traditionally, PSM has been treated as a lower level operating function which has little to do with overall corporate strategy development and implementation (Farmer, 1978; Matthews *et al.*, 2000). Much of the available literature on this subject treats PSM strategy and policy from an operational perspective (Narasimhan *et al.*, 1998; Watts *et al.*, 1995). More recent publications began to recognize the importance of PSM and its supply base in formulating and implementing corporate level strategies (Ferguson *et al.*, 1996; Kligge, 1992). One of the key problems in optimally managing the supply base is the development of a framework linking corporate level strategies to PSM strategy and their subsequent elements as described in chapter 2.3.2. This doctoral thesis focuses on the linkage between strategic suppliers' capabilities and the buying firm's strategic priorities as well as their influence on the generation of competitive advantages.

Currently, a paradigm shift is taking place concerning how suppliers are seen from a value creation perspective. In figure 16, *Watts, Kim* and *Hahn* (1995) demonstrate the existing paradigm for analyzing the buyer-supplier relationship. The former and new paradigms are presented in a two dimensional matrix using the adversarial/collaborative buyer-supplier relationships and the product/capability orientations of the buyers as key dimensions. The traditional buyer-supplier relationship in Europe and the United States might be characterized as an arm's-length, independent relationship (type 1). From this relationship perspective, a buyer and a supplier tend to see each other as adversaries competing for a larger share of the resources rather than as cooperative members of an overall supply chain or extended enterprise. This adversarial emphasis requires that firms establish bargaining power over their counterparts as shown in the market-based view focusing on monopoly rents in chapter 2.2.1. This buyer-supplier power relationship determines the relative distribution of the generated wealth between them. An adversarial relationship also tends to emphasize short-term product-based or project-based decisions. In this type of situation, buying firms tend to emphasize the current purchase price and quality of the products from the supplier rather than the long-term total costs and quality capabilities. The availability of alternative sources of supply is considered desirable and a critical element for ensuring an efficient marketplace. Consequently, a multiple sourcing strategy is frequently favored, and the supply management function tends to put its efforts into creating equivalent and reliable sources of supply. According to *Watts et al.* (1995), it is important for buying firms to understand that this short-term oriented, traditional buyer-supplier relationship is not consistent with the long-term perspective of the corporate strategy processes. As a result, PSM and its suppliers are frequently omitted from these important processes.

Figure 16: A shifting paradigm for buyer-supplier relationships

	Basis for Supplier Selection	
	Product/Commodity-based Approach	**Capability-based Approach**
Adversarial / Arm's length Approach	-Short-term/operational emphasis -Price/quality based -Multiple sourcing	-Long-term/strategic emphasis -Capability-based -Competitive focus -Multiple sourcing
Cooperative / Partner-like Approach	-Non price biased -Operational emphasis -Management assistance -Technical advice	-Strategic emphasis -Single sourcing -Continuous improvement (cost, quality, delivery, flexibility, etc.)

(Source: Adapted fromWatts *et al.*, 1995).

The developments of the last few years have stimulated discussions about a new way of looking at the buyer-supplier relationship facilitating a more active PSM role in the corporate strategy processes (Burnes, 1997; Cousins, 2002; Spekman *et al.*, 1998). The new concept is largely based on a collaborative buyer-supplier partnership rather than on an independent adversarial relationship. Especially, there is less emphasis on the power relationship because the new relationship type calls for sharing the long-term benefits and risks between the buyer and supplier. In fact, the power relationship tends to be seen more balanced, with mutual dependence becoming the key to the buyer-supplier relationship. *Watts, Kim* and *Hahn* (1995) propose that the emphasis is shifted to the long-term capability development of a single reliable supplier and away from the short-term product emphasis with multiple suppliers. This proposition is also supported by empirical results (e.g. Pearson *et al.*, 1995). This new type of relationship calls for the fact that the competitive market environments are constantly changing, and that adapting to these new environments with continuous improvements is critical for long-term survival. In this partnership-like relationship, major efforts are directed toward supporting each other in improving capabilities which impact the competitive advantages of both buyer and supplier (Monczka *et al.*, 1996). In linking PSM strategy to corporate level strategies, it is important to recognize and incorporate this new supplier relationship (Watts *et al.*, 1995). This change in buyer-supplier relationships has also been analyzed and proposed by the Industrial Marketing and Purchasing (IMP) Group for a long term (Sheth *et al.*, 1997).

The remaining two relationship types of figure 16 show the other possible buyer-supplier relationships. Type 2 recognizes the importance of supplier capabilities and tries to put emphasis on their improvement. However, the buying firm still relies on its bargaining power in dealing with its suppliers. Type 3 depicts the situation in which the buying firm views suppliers as partners. The buyer and supplier are in a more balanced power situation due to some kind of mutual dependence. However, their relationship is still based on currently supplied products or commodities (Watts *et al.*, 1995). While adversarial relationships have their justification also in today's business environment, the proposed approach shows clearly that at least the management of strategic suppliers needs to shift its focus from a product view to a more appropriate capability view when selecting strategic suppliers (Anderson *et al.*, 1999).

As a result, companies are challenged to build superior value networks by selecting the suppliers based on their capabilities in order to support buying firms' strategy implementation (Christopher *et al.*, 2000; Cousins *et al.*, 2003). In the following, the need for an alignment between corporate level strategies and PSM strategy serves as a starting point for a strategy-based selection of strategic suppliers. Therefore, the first proposition (Whetten, 1989) is stated as follows:

Proposition 1: PSM strategy and its subsequent elements need to be aligned with companies' strategic priorities.

One of PSM's key responsibilities is to ensure that buying firms have access to the capabilities and technologies in the supply base which will properly complement and support the buying firm's competitive position. To achieve these objectives, buying firms must put in place systematic processes for determining its current and future needs for external capabilities, for building a world-class supply base, and for developing the necessary access to current and emergent supply and customer markets (A.T. Kearney Inc., 2004; Monczka *et al.*, 2000a; Virolainen, 1998).

Firstly, buying firms have to determine their current and future needs. This process starts with the corporate strategies and is driving through the business units' strategies as well as functional level strategies and ends with the key supplier strategies for the supply base (Baily *et al.*, 1994; Colsman, 2000; Little, 2004; Monczka *et al.*, 2000a). Each level must ultimately ask itself: how are we optimally meeting customer requirements today, and how will we optimally meet anticipated customer requirements tomorrow? Figure 17 shows some important questions for each level.

Figure 17: Key strategic management questions

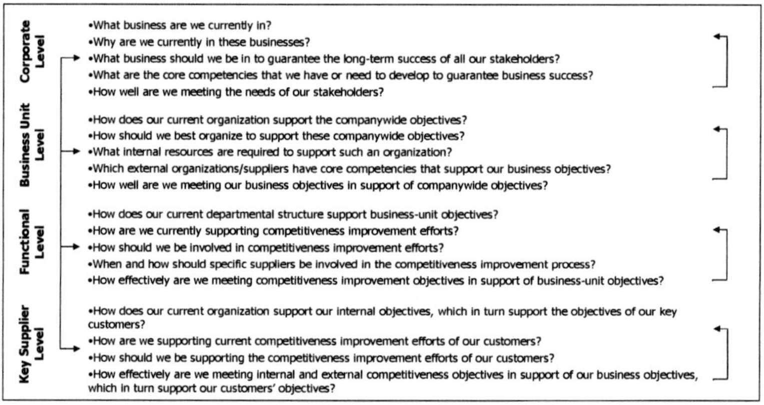

(Source: Adapted from Monczka *et al.*, 2000a)

The relationship of corporate, business unit and other functional strategies with the PSM strategy is shown by *Monczka, Handfield, Scannell, Ragatz* and *Frayer* (2000a). They propose a dynamic and iterative system which links the different strategy elements (figure 18). It shows that an integrated corporate strategy formulated at the highest executive level is the result of inputs from a number of sources such as the corporate objectives and other factors. These factors include a firm's competitive strengths, business unit and functional capabilities, market objectives, competitive pressures, customer requirements, and macro trends. The alignment between a company's strategic priorities and its external environment is an important factor of the overall company performance and defined as external fit (Pagell *et al.*, 2002). For this fit evaluation, a SWOT-analysis is a possible instrument (Westermann, 1998). Another important feature of the strategy system is the linkage, either directly or indirectly, of the PSM strategy with other functional strategies such as R&D, marketing or manufacturing also called internal fit (Pagell, 2004). The alignment of the different strategies can be seen as the strategic consensus level within a company (Morash, 2001; Pagell, 2004; Pagell *et al.*, 2002). This concept can also be adapted to the construct of sub strategies such as the supplier strategy. The higher the level of strategic consensus between the PSM strategy as functional strategy and the supplier strategy as sub element of the sourcing strategy, the better is the usage of the limited resources concerning the interaction with the suppliers.

Figure 18: Strategy level interaction system

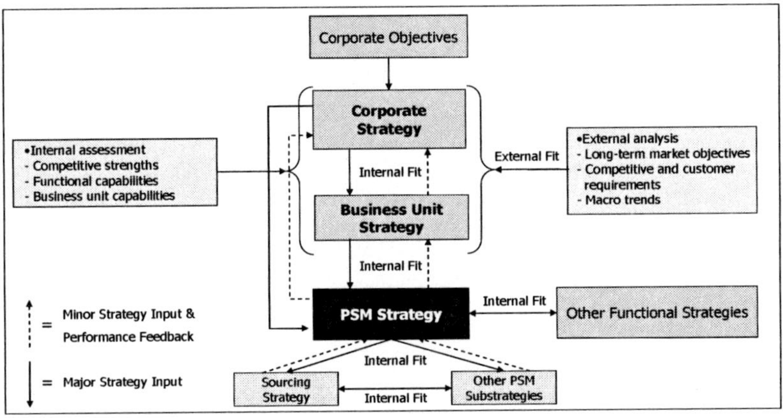

(Source: Adapted from Monczka *et al.*, 2002; Pagell *et al.*, 2002)

According to *Monczka, Handfield, Scannell, Ragatz* and *Frayer* (2000a), this linkage system recognizes the need to remove the barriers to cross-functional integration and represents a system that promotes an integrative PSM strategy development focusing on the company's limited resources towards the most relevant performance objectives. The major result of this development process is a PSM strategy that represents the primary source for the development of the commodity structure (Monczka *et al.*, 2002). The linkage system between the strategies at the different levels is the basis for the strategy-based supplier selection.

Recent studies (e.g. Jahns, 2005b; Möhrstädt *et al.*, 2001) have shown that companies' value added is often less than 50%. A significant part of companies' strategy development and implementation therefore depends on the supply base and its capabilities. Neglecting suppliers' possible contribution results in insufficient strategy realization and competitive disadvantages. In essence, PSM has the primary responsibility of linking supplier's capabilities with the internal requirements specified by corporate, business unit and other functional strategies (Fuchs *et al.*, 2000; Krause *et al.*, 1998). Therefore, PSM activities should be aligned with the nature of the corporate, business unit and functional strategies and contribute proactively to the generation of competitive advantages (Monczka *et al.*, 2002; Procurement Strategy Council, 2002; Virolainen, 1998).

Therefore, the second proposition is stated as follows:

Proposition 2: Companies select their strategic suppliers based on their suppliers' capabilities to implement their strategic priorities

Successful companies use advanced supplier selection approaches. They adapt their supplier selection to their value proposition and to the nature of their relevant markets. For example, Nokia's frequent and rapid product introductions are supported by a very flexible and responsive global supply chain and contribute to its fast sales and profit growth. In effect, Nokia has altered the playing field in the mobile phone market with rapid-response manufacturing, quick-ship logistics, and a global supply chain which links its suppliers and plants and also supports vendor-managed inventory and collaborative planning. At Nokia's best times, these capabilities have contributed to 20% margins, a 35% market share, and an average cost to make and marketing cell phones which is 18% lower than its rivals (Fisher, 1997).

The break-down of companies' strategic priorities into supplier selection priorities can start with a capability-gap analysis approach (figure 19). The performance offering model breaks down the market positioning strategy from the value chain level to the specific product and service offerings. The value proposition model defines the activities and strategic requirements at different strategy levels of the buying firm.

Figure 19: Capability-gap analysis

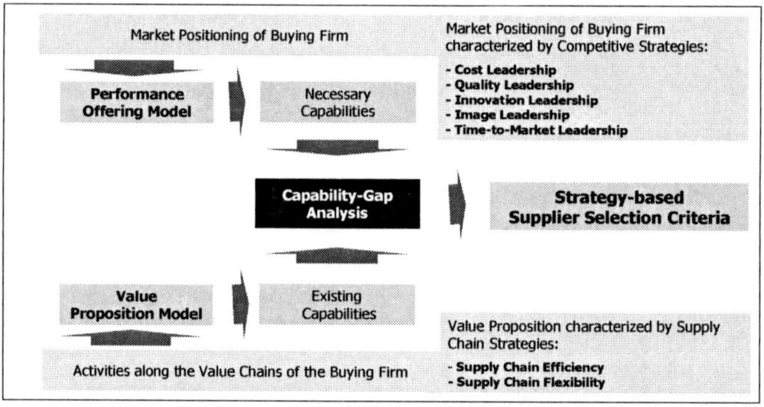

(Source: Adapted from Cravens *et al.*, 1997; Müller-Stewens *et al.*, 2003; van Hoek *et al.*, 2001)

The comparison between the performance offering model and the value proposition model results in a capability-gap analysis which defines the different capability demands on the suppliers (Carr *et al.*, 2002; Rajagopal *et al.*, 1993) and serves as the starting point for the strategy-based supplier selection.

This capability-gap looks different for each company and is changing depending on market developments and internal decisions. Nevertheless, specific strategic priorities can be examined to analyze their influence on the required capabilities of suppliers in general.

In the following, corporate level and functional level priorities are analyzed to determine how they can influence the selection of strategic suppliers (chapter 3.1). In a second step, a strategy-based selection approach is developed including selection team, process and criteria aspects (chapter 3.2). Finally, the hypotheses model and the definition of the relevant constructs for the empirical study are described (chapter 3.3).

3.1 Corporate level and functional level strategies' influence on the selection of strategic suppliers

Strategy-based supplier selection requires the integration of different corporate and functional level strategies. In the following, the possible implications from different strategy levels on the selection of strategic suppliers are discussed. Firstly, the relevance of product-/market and competitive strategies as important corporate level priorities towards markets and competitors are presented. Secondly, the influence of a company's supply chain strategy as determinant for the value proposition on corporate level is evaluated. Finally, the possible implications of different functional strategies on the supplier selection are discussed.

3.1.1 Competitive and product-/market strategies' influence on the selection of strategic suppliers

Competitive strategies focus on a company's position towards its competitors. Their priorities define how a company or business unit basically positions itself in the market. The most popular approach to distinguish basic competitive positions has been developed by *Porter* (1980) with his concept of generic strategies. Companies following a cost leadership strategy compete by providing a product or service at a standard quality for the lowest possible price. This advantage can be achieved by either performing specific value chain activities more cost efficient or designing a superior cost efficient value chain system. Furthermore, this strategy is often supported by economies of scale and the development of market power towards suppliers and competitors as suggested by the market-based view. *Porter's* (1980) second main strategic competitive choice is differentiation strategy. It includes the competition in different performance dimensions except the price of a product or service. A differentiation strategy aims at providing a unique customer value therefore generating a specific price premium. Examples of such unique customer values are Coca Cola for branding, Swatch for Design or Singapore Airlines for customer care (Müller-Stewens *et al.*, 2003). Buying firms following a differentiation strategy rely on different strategic suppliers that do not only support the differentiation features but also take care that the buying firm operates at such costs which allow generating an appropriate margin. Based on the developments of *Fuchs, Mifflin, Miller* and *Whitney* (2000), differentiation leadership strategies can be distinguished into brand leadership, technology/innovation leadership, quality leadership and time-to-market leadership from a competitive strategy perspective. The niche market strategy concept is not further discussed because different strategic implications for niche and mass markets are proposed to be neglectable compared to the differences between cost and differentiation leadership strategies.

According to *Porter* (1980), a company can follow either a cost leadership strategy or a differentiation strategy. If it follows both, it gets stuck in the middle and ends up with products that are neither cheaper nor more valuable than those of their competitors leading to below-average performance (Jahns, 2003b; Porth, 2003). This implies a position without any clear advantage. From this perspective, it seems to be impossible for a single company to follow a cost leadership and differentiation strategy at the same time. A comparable logic is described by *Towill* and *Christopher* (2002) concerning companies' agility and leanness.

The concept of hybrid strategies tries to overcome this pitfall through the integration of a time dimension. The use of hybrid strategies is closely connected with changes in the relevant customer markets. A common pattern starts with an innovative product which serves customer needs better than other products and the company earns a price premium for its differentiation advantage. After a certain time, the new product has become the new standard. Since other companies take over the standard, the price premium diminishes. Successful companies invest the earned money from the price premium to switch over to a cost strategy early enough and to outpace their competitors again (Müller-Stewens *et al.*, 2003). Empirically-based research (Gilbert *et al.*, 1987; Miller *et al.*, 1993) suggests that long-term successful companies change from cost leadership to differentiation and the other way around at specific points in time in order to outperform their competitors who stick to only one strategy. The introduced time dimension seems to offset the proposed contradiction between cost leadership and differentiation strategy but still the company sticks to only one strategy choice at a specific point in time.

Recently, the common wisdom that cost leadership and differentiation strategies cannot be implemented at the same time has been seriously challenged even more (Cummings *et al.*, 2004). Developments in information and communication technology as well as in market behavior and customer needs have influenced the concepts of organizational forms and competitive strategies. While stable market conditions and low product complexity usually favor the classical form of hierarchy to organize the value adding process from sourcing to marketing, recent developments indicate a dramatically increased destabilization of markets in terms of customer demands and an increasingly complex structure of products and services. These developments together with the possibilities of information and communication technology seem to prefer new organizational forms such as the virtual organization (Reichwald *et al.*, 2000) to optimally manage the value adding process (Picot *et al.*, 2003).

These insights are based on new models about how companies divide the various tasks necessary to produce products and services. *Rayport* and *Sviokla* (1999) describe the strategic change taking currently place in companies with a model that splits up the classical form of companies into content, context and infrastructure sectors.

Content describes which products or services a company is primarily offering while context explains how the content is provided to the customers. Infrastructure describes the conditions that are necessary for the provision of context and content. In the gas industry, for instance, the content is the gas, the gas pumps, customer cards and the integrated accounting system build the context and the overall gas supply with oil rigs, tanks and trucks stands for the infrastructure (Fleisch, 2001). Integrated companies of the industry age are thought to provide content, context and infrastructure. Due to decreasing coordination costs based on efficient information technology, the splitting of a product and the information about this specific product is easier. In some cases, the information about a product can become as critical as the actual product or service in terms of its effect on firm's profits. The splitting of information and product enables the division of content, context and infrastructure. Another model explaining the specialization trends of companies is proposed by *Hagel* and *Singer* (1999). They describe the traditional company as a bundle of three different areas: (1) customer relationship business, (2) product innovation and commercialization business, and (3) infrastructure management business.

The culture and organizational principles are the most distinguishing features for the three different areas. According to *Hagel* and *Singer* (1999), high transaction costs are the only glue keeping the three areas in a company because bundling them together into a single company forces the management to sub optimize the performance of each business in such ways that no amount of core process redesign can overcome (Hagel *et al.*, 1999). The reduction of transaction costs leads to a splitting of the traditional integrated structures into three separate areas (Fleisch, 2001).

Seen from a capability perspective, the different areas can be defined as clusters of specific capabilities. Experience has shown that the management of dissimilar capability clusters is a difficult task and that complementary capabilities develop in a better way when they are separated from a company's specialization area (Cummings *et al.*, 2004). Consequently, companies recognize the possibilities of building relationships with suppliers to manage closely complementary capabilities and their substantial advantages (Himpel, 1999). These advantages are most obvious if the objective is not the reduction of transaction costs but the improvement of the competitive position. This is often best achieved through the interactive cooperation of different organizations with distinctive but closely complementary capabilities. In this way, new forms of organizations such as logistics specialists of a logistics provider managing the whole distribution system of a supermarket chain can be developed preserving each company's distinctive capabilities (Loasby, 1998).

These considerations might help to understand the concept of mass customization (Pine, 1993). As *Porter's* (1980) idea of the stuck-in-the-middle problem has influenced the strategic thinking of generations of managers, his ideas have been put to test by some companies seeming to be able to combine cost leadership and customer closeness and other values at the same time providing customized products at low costs (Morash, 2001; van Weele, 2002). Mass customization might present a breakaway from the either-or approach by combining both cost focus/standardization and differentiation/customization within one supply chain.

Broadly defined, there exist four different types of mass customization approaches (collaborative, adaptive, cosmetic, and transparent) which are all based on the same principles of flexibility and adaptation (Gilmore *et al.*, 1997)

It seems that many companies following a mass customization strategy have their suppliers strategically aligned (Ward *et al.*, 1996). The company Dell Inc. serves as an example for collaborative mass customization. The success of Dell Inc. would not be possible without a clear supplier strategy specifying each supplier's role in the business model. While Dell Inc. is managing and constantly improving its customer relationships, its component suppliers are focusing on flexible and standardized low cost supply chain processes making Dell PCs almost unbeatable in the combination of low costs and customized systems. A successful example of cosmetic mass customization is shown by the cooperation between the designer Karl Lagerfeld and the Swedish low-cost fashion shop chain Hennes & Mauritz (H&M). Lagerfeld exclusively designed 30 clothes models for H&M providing haute couture fashion in its shops at relatively low prices (Kappeler, 2004). The customers of H&M want stylish clothes but they do not need large showrooms and sales personnel assisting them.

From this perspective, it seems as if *Porter's* (1980) paradigm of sticking either to cost leadership or differentiation has not been proved totally wrong. As the examples have shown, many business units are still focusing on cost leadership or differentiation while providing customized products at low costs together with their suppliers. The concept of mass customization requires that the different entities along the value chain agree upon the different roles they are playing and concentrate on their capabilities. This opinion is supported by other authors who identify mass customization as a clever combination of cost and differentiation capabilities within a single supply chain competing on a specific market (e.g. van Hoek *et al.*, 2001).

As companies pursue different types of corporate strategies, the role of PSM is different. Cost leadership and differentiation strategies require different types of PSM strategies and therefore also supplier strategies (Kligge, 1992). According to *van Weele* (2002), cost leadership strategies require price and costs as central questions in supplier selection. Another important criterion for supplier selection then lies not so much in the delivery time, but in the delivery reliability. Regular failure by suppliers to comply with delivery requirements results in production failures, which in turn means higher costs. In the case of differentiation the focus is on other supplier capabilities. These capabilities can be in the area of process or product innovations, quality control, lead time reduction or exchange of information (van Weele, 2002).

In general, companies are looking for strategic suppliers supporting their competitive strategies either in terms of efficiency or effectiveness avoiding unnecessary risks. Nevertheless, supply managers should keep in mind that cost leadership strategies on business unit level indicate by no means that cost factors are always the most important requirements for all strategic suppliers as has been shown in the aforementioned examples.

Product-/market strategies represent another important element of corporate level priorities influencing the selection of strategic suppliers because they define the relevant markets and products where business units are positioning themselves towards customers (Welge *et al.*, 1992). From a PSM strategy perspective, it is important whether the business unit is searching for new markets or product segments because this has relevant implications for the management and selection of the respective suppliers (Monczka *et al.*, 2000b). Figure 20 presents an extended view on nine possible product-/market strategies developed based on *Ansoff's* (1966) seminal work.

Figure 20: Extended product-/market matrix

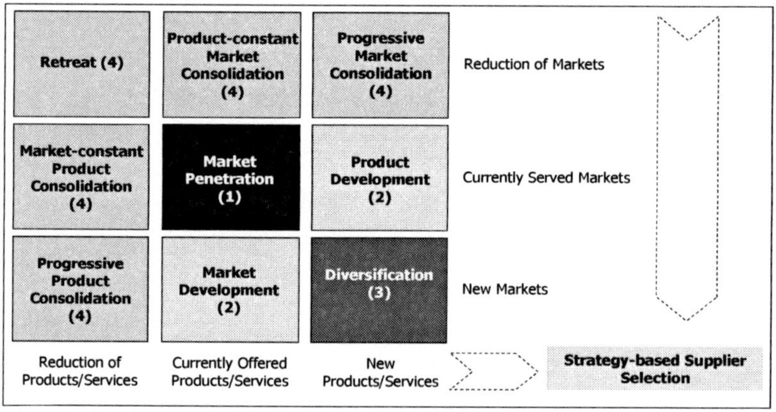

(Source: Adapted from Müller-Stewens *et al.*, 2003)

For PSM managers, it is important to know where in the matrix the product or service belongs to for which PSM has to manage the supply of raw materials, parts or systems. In the following, the different product-/market strategies are analyzed for their impact on the selection of strategic suppliers:

(1) Market penetration

(2) Product and market development

(3) Diversification

(4) Consolidation and retreat

Ad (1) Market penetration

The market penetration strategy aims at optimally using the existing volume of the company's existing market segments. The success of this strategy depends largely on the growth rate of the relevant markets. In this situation, it is likely to happen that a company cannot deliver the required quantity to its customers. If the market is satisfied, the main strategy is to win market share from the competitors resulting in large investments in sales and marketing or (short-termed) price reductions (Müller-Stewens *et al.*, 2003). In both situations, growing and stagnant markets, the role of strategic suppliers needs to be evaluated. If a company is not able to deliver the desired products or services, this can often be put down to the missing capability of strategic suppliers to grow with the same rate as the buying firm or its customer market respectively. This problem is not solely an issue of the supplier since they often do not know which growth plans the buying firm has and which capacities it should invest. It is PSM's duty to assure that the chosen supplier is able to keep up with the planned growth rates either by choosing the supplier with the necessary capabilities during the supplier selection process or by developing or assisting the most suitable supplier to achieve the desired level.

In stagnant markets, the strategic suppliers also play an important role. Especially in cases where price reductions are an important part of the product-/market strategy, suppliers should know in advance as early as possible which price reductions they are expected to deliver within a specific period of time. Once again, PSM is responsible for communicating the requirements to the existing suppliers and for taking such considerations into account when evaluating suppliers for new product or service deliveries.

Ad (2) Product and market development

Most companies cannot cover the total market for historical reasons. Therefore, it is only logical that companies extend their markets with existing products over time. This market development idea opens three new possibilities. Firstly, companies can expand to not yet served customer market segments, as has happened during the last years with cosmetic products especially for elderly people. Secondly, new geographical areas can be served as the entry of German car manufacturers with own production facilities and distribution channels in North America and most recently in China has shown. Thirdly, existing products or services can be adapted to different customers as the defense industry had to when providing satellite-based navigation systems for civilian car drivers (Müller-Stewens *et al.*, 2003).

An excellent example for the relevance of strategic suppliers when entering into new geographical areas provides the aerospace industry. Companies such as EADS are competing in a global economy but since many of their customers still have close relationships to national governments, strict local content requirements have to be fulfilled. This implies that EADS can open up new geographical markets with new airlines in South America or South-

East Asia only when its PSM managers can guarantee that the strict local content requirements will be satisfied either by local suppliers or domestic suppliers following the buying firms into new geographical markets (Moser *et al.*, 2005).

Product developments provide an almost immense variety of possibilities to expand the company's field of action. Existing products can be improved by additional features or services or new products can be developed. Especially companies competing in markets with short product life cycles rely on a steady flow of innovations to stay competitive and to reduce pressure on sales prices. In recent years, many companies realized that they can only manage to generate the required level of innovations when they are closely working together with a small but innovative part of their supply base. But suppliers are not only increasingly responsible for breakthrough innovations, the continuous improvement of products and services through suppliers become also increasingly important.

Ad (3) Diversification

Entering new markets with new products is the ultimate challenge for every company because it has the highest failure rates (Müller-Stewens *et al.*, 2003). Strategic suppliers often play a hidden role in these projects because they can either support or hinder it in many ways. For instance, if a company decides to enter into a new product market where one of its main suppliers is also competing, it may take its original business at risk. An example of this situation might be a firm buying circuit breakers from a single supplier to produce panel boards and decides to enter a new market where this supplier is also competing. If the decision is taken, PSM's role is to communicate this to the affected supplier and to identify other suppliers in advance in case the current supplier is not willing to deliver anymore.

On the other hand, PSM might explicitly focus on the selection of suppliers which can support the company's entry in new markets with new products through its specific capabilities and a strong financial background.

Ad (4) Consolidation and retreat

If a product or services of a company is thought to be in the consolidation or retreat phase, this is important information for PSM. It implies for example, that the selected supplier must be able to stay competitive even with decreasing supply volumes. This circumstance needs to be taken into account when selecting suppliers for other projects which the current supplier would also be able to deliver.

The analysis of the competitive and product-/market strategies shows that PSM needs to integrate companies' strategic priorities and analyze in detail which capabilities the suppliers need to provide in order to optimally support their implementation. The strategic priorities provide PSM managers with the relevant information to choose the best capable supplier with the most strategic value. If there is no capable supplier available they need to develop suppliers which are able to support the implementation of the chosen competitive and product-/market strategies.

3.1.2 Supply chain strategies' influence on the selection of strategic suppliers

A company's supply chain strategy serves as a good indicator for the strategic priorities of its value proposition model. Beside the market positioning towards customers and competitors, companies also have to determine which supply chain activities they want to perform themselves and where they want to rely on suppliers. Furthermore, they have to align their supply chain activities with their market positioning priorities. In the following, the implications of supply chain strategy priorities for PSM are analyzed.

An increasingly popular sentiment holds that the nature of competition in the future will not be between companies but rather between supply chains. In part, this notion might be true but a thorough analysis of companies' current business situation shows that there might be also other forms of competition in the future. For example, *Rice* and *Hoppe* (2001) propose that much of the innovation affecting the nature of competition and the supply chain of the future will relate to new and different entities that will coordinate across the supply chain. These new entities might provide unique sets of capabilities, enabled by new governance methods which work equally well for each supply chain participant. Ultimately, *Rice* and *Hoppe* (2001) envision "*...competition based on the individual company or entity and its assembled ecosystem of capabilities – but, to borrow from the Beatles, not without a little help from their friends.*"

There exist three main ways how a company can compete in the future: primarily independently against other companies, as a member within a supply chain or as the focal company of a supply chain against other supply chains (Evans *et al.*, 2001; McIvor *et al.*,

2004). *Rice* and *Hoppe* (2001) have developed three scenarios about how the future of competition beside company vs. company might look like.[25]

(1) Competing as supply chain vs. supply chain

(2) Competing on supply chain capabilities

(3) Competing on supply chain capabilities led by a channel master

Ad (1) Competing as supply chain vs. supply chain

The supply chain vs. supply chain concept simply proposes that the nature of competition will be between groups of companies formally or informally creating independent supply chains and competing as nearly one entity (Christopher *et al.*, 2000; Chung *et al.*, 2004; Kothandaraman *et al.*, 2001; Whipple *et al.*, 2000). This takes place when a supply chain is highly vertically integrated with no common suppliers shared with competitors (type 1 in figure 21). Sole-sourcing relationships and fragmented industries with no common strategic suppliers for more than one supply chain are also proposing supply chain vs. supply chain competition.

Figure 21: Possible structures of future competition between companies

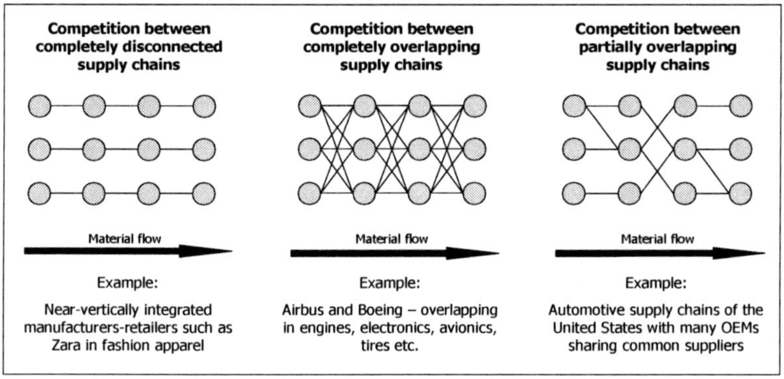

(Source: Adapted from Rice *et al.*, 2001)

In this scenario, it can happen that some companies are competing as a unique supply chain but other companies aren't. Although the competition between supply chains is currently the most popular concept for future competition, a closer examination of the supply chain vs. supply chain proposition reveals some inherent limitations which explain why it does not

[25] *Rice* and *Hope* use the term supply network for the description of some of their scenarios. As supply networks are defined in this doctoral thesis as a set of supply chains of a focal company (Harland, 1996) the term supply chain is also applicable and used for consistency reasons.

apply to all conditions (figure 22) (Christopher, 1999; Speckman *et al.*, 2002). Firstly, the presence of common or overlapping suppliers makes it difficult for a supply chain to compete as a unit for several reasons (Rice *et al.*, 2001):

(a) Common suppliers limit the ability to source unique capabilities in the form of products or services. Even if a common supplier is capable of delivering unique value to companies of different supply chains, it is still presented with a conflict of interest.

(b) Common suppliers limit the customer's ability to foster and develop unique capabilities within a particular supplier, since any investment in a supplier will also provide a benefit for competitors using the same supplier.

(c) Common suppliers make it difficult to compete without compromising other supply chain participants' business plans. The existence of common or at least overlapping suppliers complicates the task of aligning business strategies and sharing intimate business intelligence.

(d) Common suppliers inherently build a barrier to open information sharing with customers. An important issue when plans of early supplier involvement in new product or process development are implemented. Information shared by one customer with a common supplier may be inadvertently disclosed to other customers, despite the supplier's best efforts and intentions.

Another important limitation to the supply chain vs. supply chain model is the fact that suppliers often compete with customers, making true collaboration extremely difficult. An example of this situation is Siemens selling circuit breakers both to panel board original equipment manufacturers (OEM) and to internal Siemens business units which compete with these OEMs. Yet another problem is that few supply chains have a central control point which can coordinate the competitive battle against another supply chain, although this is more a practical than a theoretical problem. Finally, the risk of high sunk costs and large investments in technology dedicated to one supply chain pose other barriers to the supply chain vs. supply chain model. This is especially the case if high asset specificity is required to serve one particular supply chain.

(2) Competing on supply chain capabilities

This concept entails a single company or an entity consisting of cooperating entities, joint ventures or other legal entities competing mainly on one of two factors. These are either the cost and service capabilities or their internal supply chain. They are expanding the supply chain by utilizing and integrating the capabilities of other members upstream and downstream to offer a unique and competitive solution. The ability to integrate capabilities from other supply chain participants can often be leveraged for the generation of sustainable competitive advantages (Bozdogan *et al.*, 1998). Popular forms of integrating capabilities are joint marketing agreements, joint product development programs or collaborative initiatives such as just-in-time (JIT) (De Toni *et al.*, 2000) or vendor-managed inventory (VMI) (Disney *et al.*, 2003). The benefits of such integrating capabilities are:

(a) Benefits of one-to-one or next-tier coordination are quantifiable

(b) Data and information sharing is more immediate and useful

(c) Relationships with adjacent upstream or downstream companies are more manageable and controllable than those with more distant participants in the supply chain

(d) It might be easier to develop unique added value by working closely with one supplier, developing a unique relationship, service or product. It seems harder to do this with multiple companies in the supply chain across multiple tiers

Therefore, though it might be useful to consider various methods of coordinating across multiple tiers of the supply chain, a more practice-oriented view of the future may be one of a single company or entity competing on its own supply chain capabilities. This means competing by focusing on a company's own capabilities rather than attempting to build extended relationships with distant members of the supply chain. The important activity is integrating capabilities not only adding them since competitors can easily copy them otherwise (Monczka *et al.*, 1996). In short, the development of integrated supply chain capabilities needs to be an important part of a company's go-to-market effort. Examples of such capabilities are early supplier involvement in new product development or the adjustment of supply chain operations between two adjacent-tier companies (figure 22).

(3) Competing on supply chain capabilities led by a channel master

In this concept, the single most powerful company of a supply chain determines the terms of trade across the entire supply chain. This dominant company is often referred to as the channel master or channel captain (Christopher *et al.*, 2000). It uses its market power to coordinate processes and activities among some of its suppliers and customers. These channel masters range from being benevolent and working to provide benefits to the entire network to being entirely company-focused and transaction-oriented. In each case, they use their suppliers to influence their own cost, quality, technology, speed and responsiveness (Ragatz *et al.*, 2002). The strategic priorities of the channel master typically dictate the nature of the relationships within the supply chain and the required capabilities (Testore, 1998).

Figure 22: Examples of supply chain capability enhancements

Enhancements	Company and Initiative	Benefits
Supplier involvement	Bose Corporation's JIT II initiative gives suppliers purchasing responsibilities. Suppliers have in-plant offices and operate as Bose employees	Supplier integration enhances Bose's ability to design new products faster at lower cost and with higher quality. It lowers operating costs and improves service levels
Supplier co-location	Volkswagen's Resende, Brazil, plant is designed so that each supplier can perform on operation as vehicles move sequentially along production line	Supplier co-location improves VW's ability to reduce capital plant requirements while engaging suppliers in production
Selective and dedicated outsourcing	Apparel manufacturer and retailer Zara is almost completely integrated, outsourcing only its sewing operation	Zara is able to customize production rapidly by using local smart sewing operations. A dedicated set of suppliers lets Zara act as though vertically integrated

(Source: Adapted from Rice *et al.*, 2001)

As for the formulation of the supplier strategy and the supplier selection criteria, it is interesting to know which form of supply chain competition might outperform the others. Today, there exists no data to answer this question and most probably it rather depends on the specific industry and competitive situation. For instance, if the critical factor in a market were low costs and if there were cost advantages to having highly integrated operations, then the independent supply chain might have a distinct competitive advantage. If, on the other hand, fast cycle times or product innovations were the market's key drivers, a partially overlapping supply chain might hold a competitive advantage over the other forms. It is important to note that the proposed three main competitive scenarios are not mutually exclusive on an industry level. For example, Zara competes as a vertically integrated company based on its supply chain against a channel master (e.g. The Limited) and also against other retailers (e.g. The Gap) being part of interconnected supply chains but competing based on their own supply chain capabilities (Copacino *et al.*, 2004).

The implications of the perception and understanding of the environment in which a company competes are important for the PSM strategy and the development of the selection criteria for strategic suppliers. The PSM strategy has to answer questions such as: Does a company compete within a single supply chain, as a channel master or as a lone company solely based on its supply chain capabilities? What are the supply chain capabilities the company has and what unique set of capabilities is needed for success in the marketplace? How can the company integrate the desired capabilities – through detailed contracts or collaborative relationships? What new strategic suppliers should the company explore in order to integrate the needed capabilities? What are the trade-offs between the explicit coordination costs and the implicit opportunity costs required for the benefits of coordinating and integrating new capabilities? These are the kind of questions that companies might consider in developing the supply chain strategy and subsequently the supplier selection criteria (Rice *et al.*, 2001).

Another important aspect for companies' value proposition model is the question whether a company's supply chain is more focused on responsiveness or efficiency. *Fisher* (1997) developed a simple matrix which helps to understand the relationship between a company's market characteristics and the optimal set-up of the supply chain (figure 23). This idea is also supported by the research activities of *Lamming, Johnsen, Zheng* and *Harland* (2000) who emphasize the relevance of the product's nature for the optimal design of the supply chain. *Kaufmann* (2004a) also developed a similar analysis matrix for the supply chain.

The matrix in figure 23 shows the importance of the linkage between a company's market characteristics and its supply chain. Obviously, all four aspects can be seen as extreme positions on a continuum but they support the understanding of how important the match between value chain and market characteristics is. For example, functional products with a predictable demand include staples people can buy in a wide range of retail outlets. Many of these kinds of products satisfy basic needs which don't change much over time. This stability fosters competition and usually leads to low profit margins (Lamming *et al.*, 2000).

Figure 23: Matching supply chains with market requirements

Market Requirements / Supply Chain Nature	**Functional** Products	**Innovative** Products
Efficient Supply Chain	**MATCH**	MISMATCH
Responsive Supply Chain	MISMATCH	**MATCH**

Aspects of Demand	**Functional** (Predictable Demand)	**Innovative** (Unpredictable Demand)
Product Life Cycle	More than 2 years	3 months to 1 year
Product Variety	Low (10-20 variants per category)	High (often millions of variants per category)
Average Margin of Error in the Forecast at the Time Production is committed	10%	40-100%
Average Stock-out Rate	1-2%	10-40%
Average forced end-of-season Markdown as % of full Price	0%	10-25%
Lead Time required for made-to-order Products	6 months to 1 year	1 day to 2 weeks

(Source: Adapted from Fisher, 1997)

Many companies try to avoid this fierce price competition by competing more with innovative products that might achieve higher profit margins. Innovative products with their much more volatile demand require a completely different supply chain set-up than functional products with a more stable demand. This different kind of demand nature also influences the various functions such as physical distribution and market mediation a supply chain performs. The predictable demand of functional products makes the market mediation function a relatively easy task because a good match between supply and demand can be achieved. Therefore, companies competing on predictable markets can focus on the minimization of the physical distribution costs. In this market environment, the important flow of information is the one that occurs within the chain as suppliers, manufacturers and retailers coordinate their activities in order to meet predictable demand at the lowest cost. According to *Fisher* (1997), this is the wrong approach for innovative products with an unpredictable demand. The unpredictability for innovative products increases the risk of shortages or excess supplies. Furthermore, the potential high profit margins as well as the importance of early sales in order to achieve a high market share increase the cost of shortages. In the same way, the potential short product life cycles increases the risk of obsolescence and the cost of excess supplies. Therefore, market mediation costs dominate the physical distribution costs from a competitive point of view. In this instance, early sales numbers and other market signals are important to know and to react upon appropriately. The critical information flow starts directly in the market place and the critical decisions to be made are not about minimizing inventory and capacity costs but about where in the supply chain to position inventory and available production capacity to hedge against the unpredictable customer demand (van Hoek *et al.*, 2001).

The nature of the chosen supply chain architecture also has strong implications for the selection of strategic suppliers (van Hoek *et al.*, 2001). If innovative products are to be sold, strategic suppliers should be strong in speed and flexibility both short-term and long-term but also in joint product development or marketing support. If functional products are to be sold, selection criteria for strategic suppliers might focus more on production cost efficiency and low inventory costs through appropriate logistics systems. How ever a company builds its supply chain, the strategic suppliers should be chosen accordingly to the needed capabilities.

Although strategic PSM is proposed to be focused on the implementation of corporate level strategies, functional strategies also have significant influence on the selection priorities for strategic suppliers as they translate corporate level priorities into detailed activities.

3.1.3 Functional strategies' influence on the selection of strategic suppliers

Functional strategies add detail to a company's corporate level strategies and primarily support their implementation (Kligge, 1992; Thompson *et al.*, 1990). Recently, the traditional functional structure of many companies has been challenged by the ideas of process-based organizations. Although this development has weakened the functions' influence within the company, it does not imply that functional strategies are irrelevant for the generation of competitive advantages. Functions such as marketing, production or R&D are still the place where corporate level strategies are actually implemented (Müller-Stewens *et al.*, 2003).

The strategic priorities of these functions should be based on companies' corporate level strategies as described in the previous chapters. In practice, functions might manage their activities to optimize their own functional performance leading to suboptimal results for the overall company performance. It is PSM's key responsibility to ensure that it provides the functions with those supplier capabilities which primarily support the companies' corporate strategies such as competitive strategy or supply chain strategy. Nevertheless, the capability requirements of the functions have significant influence on the relevance of the selection criteria for strategic suppliers.

In the following, six functional strategies are analyzed concerning their interaction with PSM and their possible influence on the selection priorities for strategic suppliers:

(1) Production strategy

(2) Logistics strategy

(3) Research and development (R&D) strategy

(4) Risk management strategy

(5) Marketing strategy

(6) Financing strategy

Ad (1) Production strategy

Production strategies represent a pivotal part of companies' strategic behavior strongly influencing companies' main operations (Bozarth *et al.*, 1998b; Sun *et al.*, 2002). Cost, quality, delivery and flexibility are the most important competitive production priorities (Sun *et al.*, 2002). From a classic perspective, production depends on PSM in order to receive the right material at the right time at the right place at the right price. Based on the production schedule and the material requisitions PSM managers have to search for, negotiate with and integrate suppliers' deliveries into the production system (Bellmann, 2002). If production does not provide PSM managers with sufficient information and time to do their job professionally many needless expenses inevitably creep into the final costs of a company's products or services through premium prices, special production runs or additional transportation costs. Even more dangerous is the selection of incapable suppliers or logistics providers often leading to serious production shutdowns and interruptions in the supply chain. Production managers' main objectives are often low production costs and stable product quality through smooth and uninterrupted production processes. The potential threat of production stoppages causes many production managers to advocate an excessively large inventory of production materials (Bellmann, 2002). This often leads to much too large inventories with costs of carrying them of up to 35% of the average inventory value per year negatively affecting overall company performance (Dobler *et al.*, 1990). Therefore, production and PSM need to properly determine which inventory level is in the interest of the overall company and determine the requirements for suppliers accordingly. Furthermore, production's other strategic priorities such as quality or flexibility also have to be transformed into adequate selection criteria often creating conflicts with other functions' interests. In summary, production is often PSM's largest direct internal customer in terms of required materials and services therefore having substantial influence on the selection of suppliers. As a result, PSM managers have to carefully analyze production's strategic priorities and integrate them in accordance to company's strategic priorities.

Ad (2) Logistics strategy

The strategic priorities of logistics are closely related to production and are often opposed to those of PSM. While the logistics function primarily aims at high delivery performance requesting ideally short-term flexibility with slot size one delivered just-in-time and secondly at low transportation costs, PSM has further objectives to integrate such as low total cost of ownership for sourced products or services, global innovation identification or the reduction of capital employed (Baumgarten *et al.*, 2002; Bellmann, 2002). For example, PSM's global sourcing activities heavily influence the transportation risks and costs of inbound logistics through the increased complexity of intermodal transportation means (Stölzle *et al.*, 2002). This increases the demands on the logistics department while often aggravating its performance. Logistics and PSM managers need to jointly evaluate which strategic priorities are prioritized for the sake of the company's overall performance.

Ad (3) Research and development (R&D) strategy

The R&D function has the initial responsibility to develop new products or services and to prepare the technical specifications for a company's products or services. PSM and R&D often differ in their strategic priorities (Eger, 1998). R&D naturally tends do design conservatively, integrate the newest available materials or add unnecessary gadgets. Based on its philosophy, it is often inclined to seek the perfect product using the best equipment and procedures without complete regard for cost or timing. In such situations, R&D often develops quasi-monopolist suppliers through early specification agreements with their preferred suppliers causing a negative competitive environment (Anders, 1992). PSM managers' objectives on the other hand are focused on the performance requirements based on market expectations and cost reductions. These requirements often differ from the agreements between R&D and their preferred suppliers. For example, PSM managers often have to question whether an expensive design with a high safety factor is necessary if a less costly design with a lower safety factor will meet the customers' requirements. Why use costly chrome plates if brushed aluminium is adequate? These are possible questions from PSM to R&D (Dobler *et al.*, 1990). On the functional level there exists no proper answer to such questions. PSM managers as R&D have to find an agreement based on the company's strategic priorities often preventing unnecessary technological specificities and contributing to the reduction of costs for materials and services.

However, PSM managers can support R&D not only in terms of cost reductions but also in generating innovations and new technologies through their in-depth knowledge of the supply markets (Bratzler, 1999). Especially the identification of technological innovations is a challenge to PSM managers. Advanced companies are already aware of the strategic importance of suppliers in delivering technological innovations as the costs, complexity and rate of technology changes increase, and competition possible sources of technology globalize (Humphreys *et al.*, 2000; Stover, 1998). Especially suppliers with innovative capabilities have to be identified early and integrated accordingly into the buying firms' processes. Whatever innovations the company is currently looking for, PSM managers should be informed and even actively participate in discussions about possible new business solutions detected in the buying firm's supply base. In summary, PSM and R&D might often have contradictory objectives in terms of prices and materials used. Nevertheless, PSM can also support R&D through the integration of innovation focused selection criteria for strategic suppliers.

Ad (4) Risk management strategy

Complex and highly differentiated supply chain structures increase companies' strategic vulnerability (Krapfel *et al.*, 1991; Slack, 1983). Interruptions in supply chains, for example, can have tremendous impact on companies' financial performance (Kilgore, 2004). Risk management therefore seems a strategic necessity.

The development and implementation of a company-wide risk management strategy requires the collaboration of virtually all company functions (Jorgensen, 2005). The developed risk management strategy includes statements towards the relevant financial risks such as currency developments (Carter *et al.*, 1993) or product liabilities and operational risks from port shut-downs to corn shortages and how they should be covered (Kilgore, 2004).

The resulting strategic risk management priorities have to be reflected by PSM managers and integrated into the selection criteria for strategic suppliers. There exist several approaches (e.g. Meulbroek, 2002; Sharman, 2002; Zsidisin *et al.*, 2000) how to manage risks but generally companies can prevent, minimize, transfer or bear risks (Jahns *et al.*, 2005c). Risk managers and PSM managers have to agree on how strategic suppliers are integrated into the risk management activities such as transferring financial or operational risks to them.

Ad (5) Marketing strategy

Marketing and PSM have many fields of interaction. For example, PSM and marketing managers must wisely blend their interests in the delicate area of barter and reciprocal transactions. If satisfactory legal reciprocal transactions can be developed, they must be planned with an understanding of the total costs of reciprocity. Purchasing from customers may be a good idea, but not if it is done at the expense of other relevant strategic priorities such as quality, flexibility or even cost. In quest for sales increases, marketing managers might lose sight of the fact that sales increases not always directly result in increased profits (Dobler *et al.*, 1990). The increasing product variety is another important problem which is situated between marketing and PSM. PSM managers seek to reduce costs by sourcing for standard parts but marketing often wants to offer a wide range of variety to increase sales figures. This development leads to complexity and higher costs in sourcing, operations and inventory (Anders, 1992). The most recent battle field between marketing and PSM is on companies' marketing spend. On average, companies spend about 4 percent of their sales volume on marketing with industries having much larger percentages. A recent study by *Jahns et al.* (2005a) has shown that only a few companies already integrated PSM into marketing spend decisions.

Beside this internal relationship between marketing and PSM, marketing's strategic priorities concerning customer markets, product markets or geographical markets are of pivotal importance for PSM's supplier selection decisions. For instance, suppliers can support the entry into new geographical markets with local customer knowledge or the satisfaction of local content requirements (Moser *et al.*, 2005). Another relevant aspect is suppliers' support through their brand strengths (Bengtsson *et al.*, 2004) if marketing managers want to apply ingredient branding. Marketing's strategic priorities need to be integrated into selection decisions for strategic suppliers optimizing their strategy implementation support.

Ad (6) Financing strategy

PSM's relationship with the finance department is somewhat different from its relationships with production, R&D or marketing. The difference is based on the fact that cost determinations cannot be hidden in the supply-finance relationship as they often can in the other relationships. There are many economic factors which periodically bring about favorable and completely unexpected purchasing opportunities. For example, a supplier may have excess capacity at the moment because of the cancellation of a large order from another customer. In this situation, the supplier might sell products at prices designed to recover only the direct costs. The potential profit from such unexpected purchasing opportunities must be weighed against the potential profit from other alternative uses of the buying firm's money such as capital market investments or promotional efforts in marketing. Regardless of the obtainable price advantage, the right time to purchase from the perspective of business conditions is not always the optimal time to purchase from the viewpoint of the company's treasurer. If the PSM function makes commitments to take advantage of unusually low prices without consulting the finance department, the buying firm could find itself paying for these purchases with funds needed for other purposes. On the other hand, if the finance function does not strive diligently to make funds available for such favorable purchasing opportunities, the buying firm may later on have to pay higher prices for the same material (Dobler *et al.*, 1990; Ellram *et al.*, 1995a). Another important issue in the relationship between PSM and finance is the paying policy towards to suppliers. PSM has a direct impact on the amount of cash through its payment terms and agreements with its supply base. By this way, PSM managers can provide the buying firm with investment possibilities by intelligently negotiating payment agreements. Many companies in retailing make a considerable part of their profits due to the fact that they pay their suppliers much later than that they sell the delivered products.

The recent developments in corporate governance requirements such as the Sarbanes-Oxley Act (SOX) add another dimension to the relation between finance and PSM. Among others, the Sarbanes-Oxley Act should identify and reveal the areas of material risk (Cavinato, 2003). In its section 404, material issues come into the financial world. For example, there is no problem if a supplier goes out of business when a replacement can easily be found based on a multiple sourcing strategy. However, a single sourcing strategy becomes significant from a Sarbanes-Oxley Act perspective if this causes a potential vulnerability in the supply chain (Yuva, 2004). If the finance strategy includes the proper introduction of the Sarbanes-Oxley Act, PSM has not only to adapt the supplier selection and evaluation processes but also all other relevant PSM processes that cause a specific risk level (Chou, 2005).

Another aspect includes the financial engineering activities of the finance department. For instance, it often happens that accounts payable receives improper invoices from suppliers. Estimated costs of an improper invoice are often higher than the price itself. To reduce these unnecessary costs and increase the efficiency of accounts payable, supply managers should properly communicate the specific requirements to their suppliers.

Finally, finance strategy and PSM might have to interact in reducing fixed assets. From a financial viewpoint, fixed assets or the amount of invested capital is seen as a burden for the company because fixed assets influence the company's financial flexibility and financial performance measures (Yuva, 2004). For example, PSM can reduce capital employed by prioritizing suppliers with flexible delivery or innovative financing possibilities.

As a result, many aspects of the finance strategy have a direct impact on the PSM strategy and the management of the supply base. It is of vital importance for both functions PSM and finance that they agree on their joint strategic priorities in order to achieve and support each others objectives. Targets such as cash-to-cash cycle times, level of capital employed or requirements from finance-related issues such as the Sarbanes-Oxley Act or intelligent financing instruments (e.g. leasing vs. buying) must be known by PSM managers to optimally select their strategic suppliers and to support the overall objectives of the buying firm from a financial perspective. The currently often seen direct reporting of CPOs to CFOs is just another indicator for the increased impact of PSM on companies' overall financial situation.

Having discussed the influence of corporate strategies and functional strategies on PSM and a strategy-based supplier selection it is necessary to further analyze and discuss possible supplier selection approaches and develop an integrated framework.

3.2 A strategy-based supplier selection framework

A strategy-based supplier selection is not solely based on metrics such as cost, quality, time and flexibility but also incorporates a wider strategic perspective focusing on the capabilities of the suppliers such as process management capabilities, design capabilities, product knowledge or risk reduction capabilities into the decision-making process. As presented, a company's strategic market positioning priorities and value proposition strongly influence the determination of the required capabilities from strategic suppliers.

A strategy-based supplier selection has to tailor the selection criteria for each strategic supplier and the respective supplied commodities to the specific strategic requirements from the internal customers (Scott, 2004) as indicated in the sourcing strategy concept of chapter 2.3.2. PSM managers need to understand how their company is competing in the supply chain and what contributions they have to require from each specific strategic supplier. For example, Xerox has pursued a quality-oriented strategy for many years. More recently, the company decided to slightly change its course towards a stronger emphasis on costs which also had an impact on the supplier selection as cost had gained in importance while quality had become more a qualifier (Christopher *et al.*, 2000).

But a company's strategic priorities are not the only relevant aspect in order to select the strategically optimal suppliers. A strategy-based supplier selection framework also has to be integrated and managed accordingly. Therefore, organizational aspects, process aspects and content aspects for a strategy-based supplier selection are discussed in the following.

3.2.1 Organizational aspects of strategy-based supplier selection

Organizational aspects in strategy-based supplier selection include two main issues. Firstly, PSM's supplier selection responsibility and the integration of other functions are discussed. Secondly, the specific part of the supply base is determined where a significant allocation of PSM's resources for supplier selection is justified.

If supplier selection is the responsibility of a single function it is most likely PSM making the decision. This can happen by a single PSM manager or by a commodity management team (Johnson *et al.*, 2004). Although other functions might have more influence through processes which occur before the formal supplier selection process starts, PSM is considered to be formally responsible in most companies. With the decision authority united in a single function the decision-making process is likely to be efficient but not always effective. There is a clear advantage if suppliers have a single point of contact. This prevents misinterpretations and wrong promises from other functions. But experience shows that when supplier selection responsibility is in the hands of a single function such as PSM, the information sharing about potential suppliers and strategic priorities is often limited. This seems primarily due to unawareness and wrong assumptions about the buying firm's needs. Relevant supplier selection information, even if available, does not get passed on from other functions to PSM. Furthermore, if a single function rules the selection process it is likely that others disagree with its decisions. As a result, other functions such as R&D, production or quality management could undermine PSM's decisions by setting requirements that are against the overall interest of the buying firm, optimize their own strategies or simply bypass PSM's supplier selection decisions (Peter, 1996). For example, while R&D and production often have an interest in focusing on technology and quality measures, PSM managers more often highlight cost requirements. Additionally, marketing might be in favor of criteria such as a supplier's brand strength or its possibility to open new product markets through local content requirements while PSM has also to focus on delivery capabilities.

The integration of all relevant strategic priorities and the achievement of acceptance from other functions require a broad network within the company. Supplier selection in a cross-functional approach has the advantage that more functions can provide input and openly communicate their specific requirements. Cross-functional teams can either be project-based or permanent. Mostly, highly ranked members of various functions are involved to agree on a specific strategic supplier. Nevertheless, strategic supplier selection usually takes place in permanent cross-functional teams that are organized for specific categories or commodities (Giunipero *et al.*, 1997).

This cross-functional approach ensures that all relevant strategic priorities are considered when selecting a supplier. By using the team's cross-functional knowledge, the selected suppliers are more likely to meet the buying firm's overall strategic requirements (Laseter, 1998; Trent, 2004). At least, since all relevant functions can participate in the selection process the entire team is accountable collectively for the selection decision. Naturally, a cross-functional team decision process causes higher organizational efforts up-front in order to select the supplier. The amount of required communication adds to the overall project complexity.

Nevertheless, only a cross-functional team approach with clear responsibilities can integrate the required wide variety of strategic priorities and work effectively (Anonymous, 1988; Pearson *et al.*, 1995; Peter, 1996). Several empirical studies have shown that cross-functional supplier selection has increased but still needs further support through team-based reward systems and empowerment efforts (Giunipero *et al.*, 1997; Johnson *et al.*, 2004; Pearson *et al.*, 1995).

Constellations of strategic supplier selection teams often include PSM managers, quality managers, R&D personnel, technicians and general managers (Lemke *et al.*, 2000). How diverse supplier selection and evaluation teams can be combined shows Volvo's approach for module suppliers. Depending on the situation at hand, Volvo uses a number of different dimensions, criteria, content scopes, time horizons and methods. Therefore, experts with different knowledge from several departments are involved in the evaluation and selection decisions for module suppliers' in terms of output, processes and structures (figure 24).

PSM managers and other involved selection personnel should not only be able to assess suppliers' capabilities but also to understand the strategic priorities of their companies and to fill the identified capability-gaps. A team-based selection process is often time-consuming and costly. Therefore, this approach should only be applied to strategic suppliers that have a significant impact on a company's strategy implementation.

Figure 24: Volvo's module supplier's performance evaluation

Dimensions, Criteria and Scope	**Method**	**People Involved (Department)**
Module Quality Performance		
Function, Geometry, Looks and Noise Module Features at and after the line	Formal, Quantitative & Qualitative	Assembly Operators (Assembly) QA-Engineers (Assembly) SQA-Engineers (Logistics)
Quality Processes and Structures 1. inside module supplier, 2. on its supply side, 3. in interaction with Volvo	Semi-formal, Quantitative & Qualitative	SQA-Engineers (Logistics) Assembly Managers (Assembly)
Delivery Precision Performance		
Module Carrier on time at loading dock	Formal, Quantitative	Delivery Controllers (Logistics)
Modules in right box in carrier at line	Formal, Quantitative	Assembly Operators (Assembly)
Logistics Processes and Structures 1. inside module supplier, 2. on its supply side, 3. in interaction with Volvo	1. Semi-formal, Quantitative & Qualitative, 2.& 3. Formal, Quantitative & Qualitative	1. Delivery Controllers (Logistics) 2. Delivery Controllers (Logistics) 3. Logistics Engineer (Purchasing)
Cost Performance		
1. Module Price, 2. Logistics Costs	1. Formal, Quantitative, 2. Formal, Quantitative & Qualitative	1. Purchaser (Purchasing) 2. Logistics Engineer (Logistics)
Overall Performance		
Quality, Delivery, Cost, Management, Supply Management, Environment	Semi-formal, Quantitative	Purchaser & Purchasing Engineer (Purchasing)

(Source: Adapted from Fredriksson *et al.*, 2003)

There exist several ways to determine the suppliers that should be assessed from a strategic perspective. Generally spoken, companies organize their suppliers in supplier hierarchies, supplier portfolios or a combination. In practice, most companies prefer to work with supplier hierarchies (Christopher *et al.*, 2000).

A valuable contribution to possible forms of strategic supplier segmentation comes from *Dyer, Cho* and *Chu* (1998a). Based on their research they propose that suppliers should be analyzed strategically and then segmented into two primary groups. The first group of suppliers is providing goods or services which are necessary but not strategic inputs and the second group of suppliers is providing strategic input. They define the term strategic as high-value input that is related to the buying firm's core competence and/or provides the possibility to support the company's competitive strategy. As an example they use the Japanese auto industry where such things as transmission and engine parts, air conditioners, and instrument panels are strategic. These inputs are supplied by affiliated suppliers that customize their products and therefore support the buying firm's market positioning as proposed earlier. Most often, independent suppliers provide the non-strategic inputs such as belts, tires, and batteries with a low or even non-existing degree of customization and missing differentiation potential.
Another practice example is Siemens' supplier evaluation and selection approach (figure 25). Detailed supplier analyses primarily take place for the largest suppliers covering 80% of the total purchasing volume. Furthermore, future large volume suppliers and suppliers delivering strategic products and services are analyzed in detail. In this way, Siemens efficiently covers volume suppliers which are often selected to provide cost advantages and other suppliers which significantly can contribute to the generation of competitive advantages.

Figure 25: Siemens' supplier separation approach

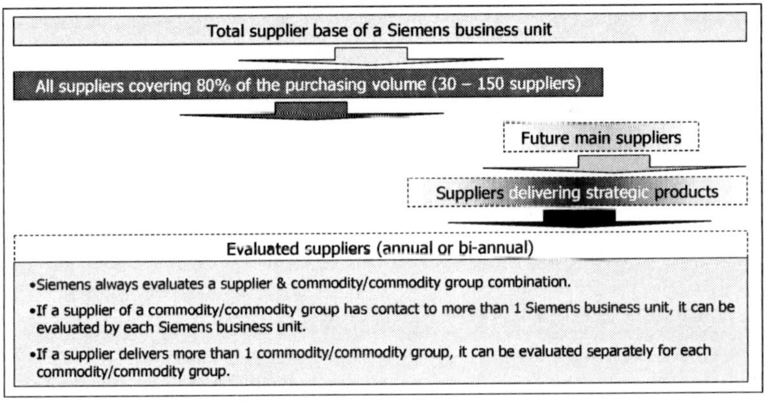

(Source: Adapted from Hildebrand *et al.*, 2000)

Nevertheless, Siemens' detailed analysis of suppliers is not based on a clear distinction between suppliers providing competitive advantages and suppliers which do not. Especially large-volume suppliers often deliver products to competitors for the same prices and conditions and therefore do not generate competitive advantages. In this case, the focus should lie on process efficiency and less on suppliers' strategic effectiveness.

Based on these insights and the concept of strategic suppliers in chapter 1.1, PSM managers might separate their supply base into suppliers that can potentially provide significant competitive advantages by generating monopoly, Ricardian, Schumpeterian or relational rents and suppliers providing products or services with no rent generation opportunities. Only with this separation the term strategic supply base is appropriate.

Recent research also supports the idea that instead of employing a "one-size-fits all" selection strategy, firms should think strategically about segmenting their supplier base in order to allocate different levels of resources to each group. Companies should allocate the most selection resources to segments that are most valuable to the company's performance and in the same sense, allocate less resources to suppliers that provide low value to the company's overall performance (Procurement Strategy Council, 2002). The most important implication for PSM managers is a structured analysis of each supplier and its capabilities based on the rent generation possibilities. As a result of this analysis, the supply base needs to be separated into a strategic part and a non-strategic part and managed accordingly.

3.2.2 Process aspects of strategy-based supplier selection

The integration of different strategic priorities and functional interests in supplier selection processes requires a systematic procedure. As the main separation of the supply base in strategic and non-strategic suppliers has already indicated, a one-size-fits-all supplier selection approach is neither efficient nor effective. According to *Kleinau* (1994), the most important criticisms concerning many currently applied supplier selection procedures are as follows:

(a) Only a few supplier selection processes concentrate on long-term, strategic aspects of buyer-supplier relationships.

(b) Many models show a low flexibility in terms of objectives since the selection of a system supplier should be based on other selection criteria than the selection of standard product supplier.

(c) All kinds of models have problems in terms of quantification and inter-personnel comparability. Especially with long-term and strategic selection objectives, qualitative criteria become more important but can hardly be quantified in an objective way.

In order to develop an appropriate selection approach for strategic suppliers, a suitable model should provide PSM managers with the opportunity to integrate selection criteria based on different dimensions and ensure that quantifiable and non-quantifiable criteria can be integrated. *Kleinau* (1994) proposes the application of a common assessment scale based on an interval-scale dimension. In order to ensure the integration of quantifiable and non-quantifiable, the transformation of ordinal-scale criteria into interval-scale criteria is necessary. As a first positive result of such an approach, PSM managers can process data with statistical methods.

A strategy-based selection of suppliers also involves the weighting of the different criteria in order to select the supplier(s) that best suit the strategic requirements of the buying firm. An analysis of general supplier evaluation and selection models is presented by *Glantschnig* (1994). In strategic assessments, different capabilities are considered as the overall performance of the supplier and no specific product or service comes into focus (Dulmin *et al.*, 2003). Mostly, a choice involves a trade-off among the different capabilities of the assessed potential strategic suppliers. Therefore, the weighting of the relevant factors may have a substantial impact on the selection results and the final choice of suppliers (Petroni *et al.*, 2000). PSM managers and their colleagues from the selection team must analyze and document the relevance of each criterion based on the company's strategic priorities, converting the more or less instinctive qualitative factors into empirical measures. It might be even impossible to get purely quantitative-based scores for the various criteria because the selection of (new) strategic suppliers might include facts that are hard to quantify such as new technological solutions, risk management capabilities or market entry support. Since many strategic suppliers are selected in order to develop new products or services or improve production processes for which no databases exist, the estimation or measurement of the required supplier capabilities is difficult.

In the next sections, the focus lies on the problem to optimally weight the various selection criteria based on companies' strategic requirements.

Several techniques used for supplier selection assign importance weights to selection criteria in a subjective and/or arbitrary manner. It is important to realize that as the complexity of the strategies pursued grows, it is increasingly difficult to assign a consistent set of values. Many researchers have proposed new approaches based on different statistical techniques to overcome these obstacles (e.g. Narasimhan *et al.*, 2001b; Talluri *et al.*, 2004).

The main motivation for the development of new approaches comes directly from practice problems in supplier selection. Firstly, most supplier selection approaches used in practice are based on simple, weighted scoring methods primarily relying on subjective judgments and opinions of supply professionals and other involved people in the supplier selection process (Knapp *et al.*, 2000). Although this approach has its advantages in form of time efficiency or use of experience, it also has some limitations in such a way that the weightings for the various supplier selection criteria are often more or less arbitrarily set. As a result, the final choice of a strategic supplier might be heavily dependent on the assignment of the weightings of the selection criteria. Secondly, many supplier selections are currently done by groups of different functions and in changing combinations. Although it might not be difficult to achieve concurrence on the two or three most important selection criteria, it is often hard to reach consensus beyond those first few factors. In practice, mergers, resignations, job reassignments and other incidents cause changing group compositions for the selection of strategic suppliers. This leads to constantly necessary new discussions about the relevance of the different criteria. A methodology allowing a more objective weighting process might avoid costly and time-consuming discussions (Narasimhan *et al.*, 2001b).

Until today, none of the many developed concepts has been widely accepted by practice beside simple adaptations of the weighted point method (Stueland, 2004). Although some concepts and ideas have provided important insights into certain problem areas, none of the new approaches has made its way into the board rooms. It might be possible, that combinations of aspects of different concepts support PSM managers and their colleagues in overcoming some of the discussed obstacles. Since each company has its own strategic agenda, team constellations and political problems, it is up to the PSM managers to choose and combine the different possible selection approaches to fit their needs. In the following, the different characteristics of some of the most applied and discussed approaches[26] are presented. These approaches are:

(1) Weighted point method

(2) Data envelopment analysis

(3) Analytical hierarchy process

(4) Analytical network process

---

[26] There exists a wide range of methods for supplier evaluation and selection. Beside the methods discussed, many other concepts have been developed such as categorical methods, cost-ratio methods, linear programming models, clustering methods on performance factors, matrix methods, multi-objective programming methods, cost ratio methods, total cost of ownership methods, human judgment methods or interpretive structural methods (e.g. Narasimhan *et al.*, 2001b; Petroni *et al.*, 2000; Sarkis *et al.*, 2002). Nevertheless, the presented approaches are perceived to be the most appropriate for a strategy-based supplier selection.

The following discussion should help to understand some important weighting aspects of supplier selection procedures and the possibilities to achieve more objectivity in selection criteria weighting.

Ad (1) Weighted point method

The primary issue in weighted point methods is identifying appropriate weights in calculating a composite index for supplier performance. The weighted point methods consider selection criteria that are weighted by supply professionals and other members of cross-functional supplier selection teams. The weight for each selection criteria is then multiplied by a performance score that is also assigned by the selection team. Finally, the products of weighting and performance scores are totaled to determine a final rating for each assessed supplier (Petroni *et al.*, 2000; Riffner *et al.*, 2001). A good example for such a weighted point method is the value benefit analysis (Gomez *et al.*, 1997).

The problem with this method might be the absolute compensation between the various selection criteria. This means that a good performance on one factor can balance a poor performance on another. In practice, some selection criteria are sometimes sufficiently relevant to refuse any compensation (non-compensatory logic) or partial compensation (partially compensatory logic) (Dulmin *et al.*, 2003).

Through its understandable process it is one of the most common approaches in supplier selection practice (Chen *et al.*, 2004; Schmitz *et al.*, 2003). Contrary to many other concepts, the weighted point method allows to deal with uncertainty in decision making deriving from incomplete and qualitative data as well as unstructured supply situations (Dulmin *et al.*, 2003).

Ad (2) Data envelopment analysis

The data envelopment analysis (DEA) is a non-parametric multi-factor productivity analysis method which evaluates the relative efficiencies of a homogenous set of decision-making units (suppliers) (Talluri *et al.*, 2004). Although the focus usually is on input factors both input and output factors can be integrated. From the weighting perspective, this assessment concept is interesting because DEA does not require an a priori assignment of weights to the criteria used in the selection process (Narasimhan *et al.*, 2001b). This allows the selection team to concentrate on the definition of the input and output dimensions. The DEA process finally results in an efficiency score. Efficiency is defined as the ratio of weighted output to weighted input. Therefore, if a decision-making unit (supplier) has an efficiency score of one it is considered to be efficient. The mathematical model of the DEA is run n times, where n represents the number of decision-making units, and determines the efficiency scores of all suppliers. With this process, each decision-making unit is allowed to select the optimal weights maximizing its efficiency. At the same time, the efficiencies of all other suppliers in the set are prevented from exceeding a value of one when evaluated with these weights (Narasimhan *et al.*, 2001b).

In order to execute a DEA, the supplier selection team has to define and measure the context-specific supplier capabilities (input factors) which may include technical, managerial, operational, or dynamic capabilities. In the same way, the supplier selection team has to define and measure the required performance dimensions such as quality or cost (output factors) based on the requirements from the company's strategic priorities (Talluri *et al.*, 2004). With this process, valuable information can be gained about which suppliers fit best for a specific sourcing strategy and which suppliers might be worth integrating into a supplier development program. As a result, the DEA score becomes a substitute for the overall fit to the requirements of a PSM strategy. It can also provide an alternative to assess a considerable number of suppliers in a time and cost efficient way compared to supplier on-site audits. But its main advantage is the possible allowance for complete weight flexibility while weight restrictions can also be set (Narasimhan *et al.*, 2001b). This aspect makes it possible to pay more attention to the optimal definition of needed supplier capabilities (input factor) and the pursued strategic priorities (output factor).

Ad (3) Analytical hierarchy process

The analytical hierarchy process (AHP) (Saaty, 1990) is a popular approach for supplier selection and is a robust technique which allows determining criteria preferences for selection purposes, quantifying preferences, and aggregating them across different criteria. AHP is especially suitable in an environment where subjective assessments and judgments have to use in comparing factors. AHP structures a supplier selection problem in the form of a hierarchy to capture the main issues of the problem and derives ratio scales to integrate the perceptions and purposes into a synthesis. In this hierarchical structure, all the elements in a level are compared in pairs with respect to the elements in the level above, and paired comparisons are used to determine judgments. Subsequently, the synthesis of judgments is obtained as a result of hierarchic decomposition in order to find the best decision. Usually, the discussed hierarchy in supplier selection has three levels: the objectives, the selection criteria, and the alternatives (suppliers). AHP requires the selection team to develop a hierarchical structure of the factors in the given assessment problem and to provide judgments about the relative importance of each of these factors and ultimately to specify a preference for each decision alternative (supplier) with respect to each selection criteria (Bhutta *et al.*, 2002).

But there also exist some problems with this approach. For example, AHP only considers one-way hierarchical relationships among selection criteria and is only fully compensatory. This does not integrate the many possible relationships between or within the different groups of factors. Therefore, a decision maker may categorize selection criteria into flexibility, innovation, price, and risk management. A supplier may then be rated on each of these selection criteria separately and be aggregated to result in an overall performance score. AHP does not explicitly reflect the possible interactions between the various selection criteria as flexibility and innovation may have a substantial impact on risk management. Additionally, the analytical hierarchy process is rather inflexible in terms of integrating or deleting alternatives in a set of criteria. As a result, it is difficult to flexibly change the number of assessed suppliers during the ranking process. Finally, all judgments of selection criteria are always relative, therefore no absolute measures can be integrated in the supplier selection process (Bhutta *et al.*, 2002; Sarkis *et al.*, 2002). As a result, the AHP provides some new insights about how the weighting of the various selection criteria may be done in a more structured. Nevertheless, it still relies solely on the selection team's judgments.

Ad (4) Analytical network process

The analytical network process (ANP) is a generalized form of the more popular multi-attribute decision-making tool [27] analytical hierarchy process (AHP). In contrast to the weighted point methods and many other models, it explicitly integrates the interrelationships between the different selection criteria through pair-wise comparisons (Glantschnig, 1994). Like AHP, ANP is also partly based on hierarchies of clusters, factors, and sub factors etc. Its main advantage is the support of the supplier selection team in comprehending the different relationships and importance. It prevents decision makers from loosing the overview and relatedness of the different selection criteria as psychological and cognitive literature states when comparing more than 7+/-2 factors (Sarkis *et al.*, 2002). This is why ANP's pair-wise comparison between each two selection criteria allows focusing on the relevance of the respective criteria fulfilling a requirement from the PSM strategy. For example, a question could be how much more important is the selection criteria strategic flexibility when compared to the selection criteria risk management for the implementation of the buying

27 An interactive software for decsion analysis is available at http://www.hipre.hut.fi (Sarkis *et al.*, 2002).

firm's quality leadership strategy. ANP can also integrate temporally based objectives such as planning horizons or product life cycle stages.

Once the selection criteria are weighted, ANP requires the selection team also to compare the different suppliers in pair-wise comparison for each of the defined criteria. This is a rather time consuming process and seems not applicable in practice when assessing a considerable number of suppliers. As a result, ANP might become quite complex as the number of selection criteria, suppliers, and possible relationships between criteria increases (Sarkis *et al.*, 2002). If the number of supplier and criteria remains at a low level or if the supplier selection decision is of pivotal importance, the ANP might be fully executed. In this case, it provides the selection team with the possibility of sensitivity analyses. For instance, the relative importance of criteria can be changed and the impact on the suppliers' performance scores can be analyzed. This might be helpful if a small number of suppliers have almost the same (high) performance score but the number of selected suppliers needs to be further reduced.

Evaluated from a weighting perspective, the proposed method provides some important insights and processes on how to improve the weighting of the numerous selection criteria. The pair-wise comparison might be usefully applied to the weighting of the main selection criteria and maybe even in comparison with the potential suppliers. Most probably, a full application of the analytical network process with the additional analysis of each single selection criteria consumes too much time resources from the buying firm without really improving the selection process for strategic suppliers.

The four presented approaches have shown different ways how to handle the criteria weighting problem. There also exist approaches (e.g. Cook *et al.*, 1992; Turner, 1988) that are much more driven by objective weighting concepts trying to exclude the human being from this problem. From a strategy-based supplier selection perspective this does not make sense because strategic requirements cannot be integrated in this way. Therefore, these kinds of approaches are not discussed in detail.

In summary, the weighting problem is one of the important issues to be solved in strategic supplier selection. As the above discussion has shown, the weighting of the selection criteria has a severe impact on the determination of the optimal strategic supplier. Theory has developed some complex approaches (e.g. de Boer *et al.*, 1998; Dulmin *et al.*, 2003; Kwong, 2002) to ensure that all relevant aspects are integrated from objective weighting to compensatory logic and high-dimensional sensitivity analyses. In practice, the success of a supplier selection approach heavily relies on the acceptance through the selection team and the potential strategic suppliers. This implies that the chosen method should balance easy understanding and methodological correctness.

In accordance with the insight that strategic supplier selection is increasingly applied in cross-functional teams, an appropriate weighting definition process seems important. For this reason, the logic of ANP might be useful to compare the selection criteria weighting of the different members of the cross-functional team to achieve a common consensus. Based on the results the weighted point method can be applied (Janker, 2004). As the development process of the selection criteria weighting is not necessarily communicated to the suppliers, the weighting setting approach can be complex to a certain extent. The compensatory logic issue might be resolved by defining minimum scores for each selection criteria. This might prevent the disadvantages of the weighted point method's absolute compensation.

The weighting itself might be openly communicated to the suppliers and represent the buying firm's current strategic requirements (Pearson *et al.*, 1995; Stueland, 2004).

3.2.3 Content aspects of strategy-based supplier selection

A study conducted by the Cranfield School of Management has shown that the strategies and intentions of suppliers are getting more important since their impact on the performance of the buying firm is increasing (Lemke *et al.*, 2000). This development starts to influence companies' supplier selection systems because it is not possible to select and evaluate an innovative supplier accurately only based on price, quality and delivery reliability. The 'human' or 'soft' factor becomes an accepted category in supplier selection systems (Lemke *et al.*, 2000).

Supplier analysis usually happens in two different situations. Firstly, the selection of existing and potential suppliers is targeted at a sourcing object which has already been sourced and whose requirements have not changed. The main difficulty in this kind of analysis is the fact that the available comparison data from existing and potential suppliers are completely different. While the buying firm has a lot of data concerning the performance of the existing suppliers there is only few and insecure information about the possible performance of a potential supplier. Secondly, the selection of existing and potential suppliers is targeted at a sourcing object which has not been sourced yet or whose requirements have changed dramatically. In this case, there exists no database to compare the performance or even worse, the PSM managers do not even know how to measure the requirements for the sourcing object. In both cases, relevant information comparable for existing and potential supplier needs to be defined and analyzed (Folker, 1993).

The first systematic analysis of determinants influencing the selection of suppliers has been executed by *Dickson* (1966). The following years were determined by the search for economic and quantifiable criteria and the evaluation of multi-variable decision alternatives. During the last decades, the supplier selection was mostly determined by three factors price, quality and delivery reliability (Kleinau, 1994). More recently, several researchers have developed broader supplier selection and evaluation approaches.

For example, *Ulaga* (2003) has done a survey based on in-depth interviews with 10 purchasing managers in manufacturing companies asking the respondents how suppliers create value for their organizations. As a result, eight dimensions for the selection of value creating suppliers emerged (figure 26):

(1) Product quality

(2) Service support

(3) Delivery

(4) Supplier know-how

(5) Time-to-market

(6) Personal interaction

(7) Direct product costs (price)

(8) Process costs

Ad (1) Product quality

Product quality is defined as the extent to which the supplier's product or service meets the customer's specifications with the key drivers performance reliability and consistency over time. It is also the most important reason of all eight aspects for manufacturers to maintain relationships with suppliers. In today's economy, it has become increasingly difficult for suppliers to differentiate themselves from competition merely on the basis of product quality. Nowadays, quality is a given, and suppliers must meet quality standards to be included in the potential supplier short-list. According to his survey results, companies mainly search for technical performance and reliability when referring to product quality. The supplier's products are expected to meet a set of technical specifications within certain tolerance levels.

Figure 26: Relationship value drivers

(Source: Adapted from Ulaga, 2003)

Ad (2) Service support

Service support can be viewed as another key dimension of *Ulaga's* relationship value. Besides product-related services, suppliers can create value in the two other main service support areas which are customer information and outsourcing of activities. Product-related services especially include product warranties and availability of spare parts. Customer information has other facets. Firstly, manufacturers expect to get hold of suppliers whenever needed. In addition to this supplier availability, customers need to receive appropriate information in a reasonable time window. The third component within the service support dimension is the possibility of outsourcing tasks to suppliers such as assembly, design work, and product testing. For example, suppliers add value by doing the assembly through three main factors. Firstly, suppliers can create value for their customers through consolidating the supply base. This means that suppliers can reduce the number of outside companies customer need to coordinate by delivering integrated systems as opposed to single parts. Secondly, synchronizing both the supplier's and the customer's production schedules allow to deliver parts in a sequenced manner and to reduce inventories. Thirdly, the outsourcing of assembly tasks to the supplier represents a last benefit for manufacturers by liberating plant space which can be allocated to other activities in growth phases.

Ad (3) Delivery

Ulaga (2003) identified delivery performance as the third dimension of relationship value. He distinguishes three aspects of delivery performance: on-time delivery (with just-in-time delivery), delivery flexibility, and accuracy of delivery. First of all, suppliers are expected to deliver on the agreed time. Depending on the industry, this can be a time window of weeks, days, hours, or even minutes. Especially just-in-time and just-in-sequence concepts are a challenge for suppliers. Secondly, manufacturers expect their suppliers to adjust to delivery requirement changes. Such changes in delivery schedules may occur due to spikes in demand or changes in the production mix. Lastly, delivering the right parts in the correct amount seems also important.

Ad (4) Supplier know-how

Supplier know-how represents the fourth dimension of relationship value. Suppliers may have a specific expertise which is not available within the customer's organization. This increasingly important dimension encompasses several aspects. Firstly, the supplier's knowledge of the supply market provides an opportunity to present the customer with new sourcing alternatives. Secondly, a supplier adds value in assisting the manufacturer in the improvement of existing products and processes – both in terms of functionality and costs. Finally, a supplier may assist the customer in developing new products.

For instance, a supplier which has a lot of experience with a product knows how to make parts efficiently and effectively. A supplier having produced a specific part for years can make a change at maybe a third of the cost of a new supplier, and they can bring in their own experience and suggestions. Relationship value can also be created when suppliers are involved early in new product development. They are brought in as experts to suggest solutions and take cost out of the product or service right up front.

Ad (5) Time-to-market

Another relationship value dimension is a supplier's capability to reduce time-to-market. Suppliers add specifically value through accelerating design work, developing prototypes faster than competitors, and speeding up product testing and ramp-up processes. Today, speed and time-to-market have become strategic imperatives in designing and managing supply chains (Fine, 1998). Dell's direct business model might illustrate at best this trend toward shorter cycle times. The company carries inventories of only 11 days on average and delivers its PCs built-to-order within 5-6 days. At Dell, suppliers are treated as in-house partners and replenishment needs are shared in real time. Especially when new products are launched, strategic suppliers station their engineers in Dell's plants to fix design flaws in real time (Magretta, 1998).

Overall, manufacturers turn to their suppliers in different areas to reduce overall cycle times.

Ad (6) Personal interaction

Personal relationships are part of the relational exchange and buyers consider personal relationships as one important aspect of purchasing (Dwyer *et al.*, 2002). *Ulaga's* (2003) study shows a strong diversity of opinions held about the potential value creation through personal interaction. Respondents strongly supporting personal interaction as source of value creation see the most important benefits in improved communication, more effective and efficient problem resolution, and a better understanding of each partner's goal. This research area is still in its infancy and an integrated research approach is recommended since issues such as trust, communication and others have a strong history in others than business studies. Especially for managers, it might be interesting to know exactly the cause and effect relations of the different elements of the personal interaction construct (Hines, 1996).

The development of interpersonal ties leads to a number of benefits. Communication between both parties is enhanced because problems can be addressed more easily. According to *Ulaga* (2003), personal interactions should be developed at all levels of the organization, from the sales representative to the supplier's president.

There exists also the contrary opinion because too personal relationships with a supplier can affect the perception of supplier by a purchasing manager and lead to sourcing decisions that negatively affect the competitive position of the buying firm. A remedy for this problem might be transparent buying processes and decisions by cross-functional teams (van Weele *et al.*, 1996).

Ad (7) Direct product costs (price)

In his study *Ulaga* (2003) identifies direct product costs as an important aspect when selecting suppliers. Most manufacturers focus above all on direct product costs, defined as the actual price charged by the supplier for the main product sold and the sacrifice most easily identified by purchasing managers (Cannon *et al.*, 2001). A supplier's products may be priced below, above, or at competition level. Therefore, a supplier has three possibilities to price its products: at the lowest price, at the market price, or at a reasonable price. Most often, suppliers are expected to commit to annual price reductions within long-term contracts. They need to continuously identify new ways of decreasing costs and passing savings on to the buying firms. Even a collaborative relationship may protect a supplier only from competition in the way that the customer may work with the supplier to reach a competitive price level if the supplier is at a price disadvantage. Finally, switching costs represent an important factor when evaluating a supplier's direct product costs.

Ad (8) Process costs

Companies collaborate in relationships not only to get price reductions but also to achieve improvements in overall operations. According to *Ulaga's* (2003) study, several directions of value creation through cost reductions can be identified. The mostly mentioned cost categories are acquisition and operations costs. In the first category transportation costs, inventory management costs, order-handling costs, and costs related to incoming inspections occur. Operations costs as the secondary category are not seen as a primary source where suppliers can actually add value through cost reductions. Nevertheless, the most important drivers seem downtime costs, costs for tooling, warranty costs, and costs related to differences in product yields in the transformation process. As a result of the study, the distinction between the three cost categories direct product costs, acquisition costs, and operations costs seems to be difficult for PSM managers. The difficulties PSM managers find in differentiating these cost elements mainly refer to the absence of adequate information systems relating the cost categories to specific parts purchased.

Based on the study results, *Ulaga* (2003) proposes a way of profiling an existing supplier relationship and benchmark it against an alternative supplier. In his example, suppliers A and B are compared against each other (figure 27). Alternatively, a specific supplier profile may be compared with an expected or ideal profile. Firstly, suppliers are evaluated on each value-creating dimension. If needed, each dimension may be further broken down into its specific subcategories such as on-time delivery, delivery flexibility, and accuracy of delivery for the category delivery. *Ulaga* (2003) suggests to use a score within a range of 1 for very weak to 7 for very strong. Figure 27 shows that supplier A scores high on quality, service support, and delivery. In addition, supplier A offers a low purchasing price. In turn, the company does not perform well on time-to-market, supplier know-how, and personal interaction. PSM managers perceive that supplier A does not really add value by continually driving down other costs in the relationship such as operations costs. Supplier B or a potentially ideal supplier profile has a very different profile. It scores well on the value-adding dimensions of product quality, service support, delivery, time-to-market, supplier know-how, and personal interaction but is low on price (Ulaga, 2003).

Figure 27: Relationship value profiles

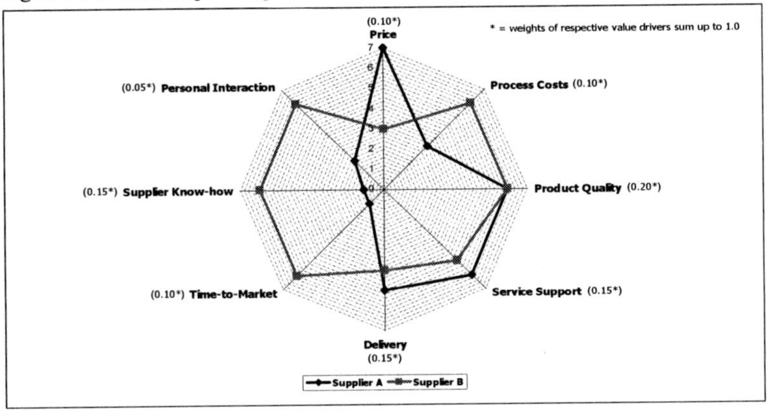

(Source: Adapted from Ulaga, 2003)

To take the supplier assessment a step further, PSM managers may attribute weights to the eight identified value dimensions. By attributing a relative importance to each value driver, an overall value score of 4.2 for suppler A and an overall score of 5.25 for supplier B results. From this perspective, supplier B would be considered to deliver more value than supplier A (Ulaga, 2003). To sum it up, *Ulaga's* paper is an important step in the further development of broader defined supplier selection approaches. His conceptualization of eight major supplier selection criteria is an important starting point for the development of a strategy-based supplier selection approach. Unfortunately, *Ulaga's* research included only 10 PSM managers not providing a representative sample. Therefore, his eight supplier selection criteria need to be empirically verified. For this doctoral thesis, *Ulaga's* research serves as a starting point for the systematic development of a broad set of strategic supplier selection criteria.

Krause and *Handfield* (1999) have developed an integrated approach to supplier development which integrates the supplier selection process in its first stage (figure 28). This first stage is seen as the necessary prerequisite for a successful supplier development and also provides valuable insights in the formulation of a supplier strategy. The first step of stage 1 identifies a need for improved supplier performance either because of customer demands or the need for global expansion. Cross-functional executive input is important at this first stage of the process because here the objective is to identify the overall business needs in terms of cost reduction goals, technology road maps, or global market expansion plans. Moreover, these needs should be explicitly identified and aligned with customer requirements and new product development targets.

Figure 28: Supplier selection as prerequisite of successful supplier development

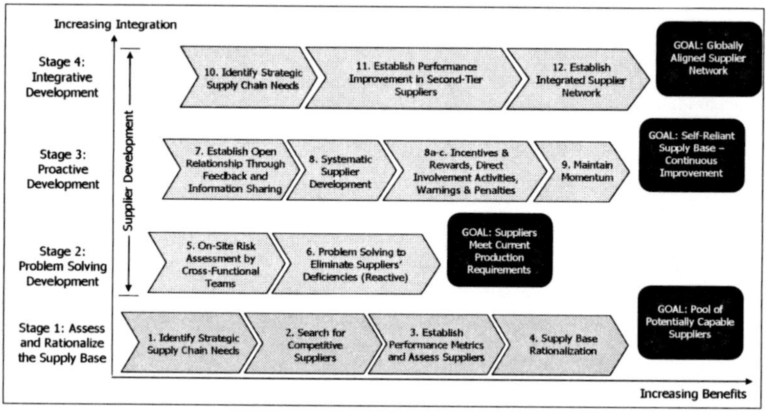

(Source: Adapted from Krause *et al.*, 1999)

The presented influence of competitive and supply chain strategies on corporate level and various functional strategies corresponds with this first step. Step two involves the worldwide search for competitive suppliers and bears in mind the criteria identified in step one. Next, the supply base is assessed in step three. As a result of step three, suppliers which are clearly not capable of meeting the organization's needs are eliminated and the supply base is optimized in step four. The outcome of this strategy is a pool of suppliers that are potentially capable of meeting the PSM organization's need for capabilities (Krause *et al.*, 1999).

As a first conclusion, *Krause* and *Handfield* highlight in step 1-4 of the first stage the importance of customer requirements for the definition of the PSM strategy and the supplier selection criteria. Additionally, they emphasize the importance of performance measurement systems which are not only focused on the past performance of suppliers in the areas of cost, time and quality (Krause *et al.*, 1999). Their concept supports the proposed alignment between corporate strategies and supplier selection. Therefore, this doctoral thesis relies also on these authors to justify the hypothesized influences of competitive and supply chain strategies on the relevance of supplier selection criteria in the empirical research model developed in the next chapter.

Kaufmann (2004b; 2006) has developed a measuring approach for strategic supplier relationships (figure 29). He argues that relationships with strategic suppliers are multifaceted and require therefore a multidimensional measurement approach.

Figure 29: X-BSC development process

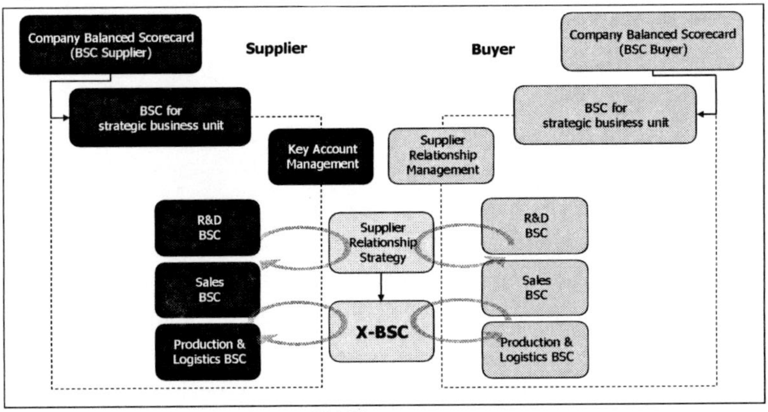

(Source: Adapted from Kaufmann, 2004b).

He describes in his concept X-BSC (Cross-Balanced Scorecard) how the principles of the balanced scorecard (BSC) can be applied to strategic supplier relationships. The X-BSC concept tries to avoid the shortcomings of classical supplier selection and evaluation systems such as the one-size-fits-all system or the transaction focus instead of a relationship focus. In addition, most traditional supplier selection and evaluation systems are backward-looking, after-the-fact measurement tools and miss a forward-looking, strategy-focused performance management concept. The X-BSC concept focuses on a buyer-supplier relationship as a whole, the future development of the relationship, and the fast implementation of jointly formulated relationship strategies. The basic idea of the X-BSC is therefore to use the BSC approach not only to implement strategies formulated within a company's boundaries but also to develop BSCs for each relationship strategy with strategic suppliers. The X-BSC translates the specific supplier relationship strategy as a whole into a set of some 20 measurable objectives in four dimensions: financial, customer, B2B, and learning & growth perspective. The X-BSC should create commitment as it becomes possible to hold the buying team and the selling team accountable for achieving the objectives of the buyer-supplier relationship.

The most important prerequisite for a successful X-BSC is the existence of a clearly articulated strategy for each supplier-buyer relationship. This usually requires the input from various departments from both sides (Kaufmann, 2004b).

It is important that buyers and suppliers do not confuse standardizing the cross-score carding process with using overly generic performance measures for each and every supplier relationship. The objectives, the performance indicators/ratios used, and the target values of these may well vary among the different strategic buyer-supplier relationships to some extent but are based on a standardized process.

This individualistic approach is quite promising because the X-BSC is planned, built, and run by the buying firm and each supplier. The supplier is more motivated for continuous improvement because it can influence the set targets. In contrast, the ratings of traditional supplier selection and evaluation systems and rules for supplier awards are often not fully understood by all suppliers leading to frustrations in many supplier-buyer relationships (Kaufmann, 2004b). Similar to Kaufmann's approach, a strategy-based supplier selection approach needs to be multidimensional, forward-looking and strategy-focused in order to result in successful buyer-supplier relationships and generate competitive advantages.

Arnold and *Eßig* (2003) developed a supplier valuing concept. This approach is based on a supplier lifetime value model (Arnold, 2004) including the insight that different economic effects on various impact levels have to be considered when developing an appropriate supplier selection concept. The developed approach mainly consists of the three systems supplier lifetime value, target supplier lifetime value and supplier valuing as integration tool (figure 30).

Figure 30: Supplier valuing concept.

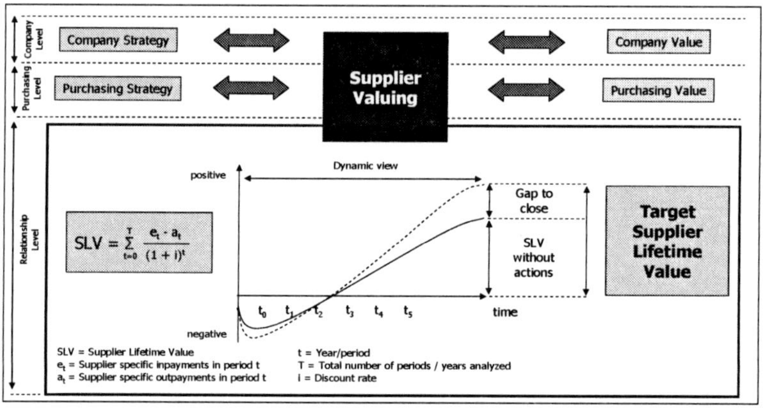

(Source: Adapted from Arnold *et al.*, 2005)

The target supplier lifetime value (TSLV) is the long-term measure of a buyer-supplier relationship's success. It can be used as a performance indicator for suppliers. By analogy with the gap analysis concept, the current supplier lifetime value (SLV) and the target supplier lifetime value (TSLV) should be compared and forecasted. Necessary actions have to be taken if a gap is identified or forecasted. The necessary actions should be initiated by the superior tool of supplier valuing. Supplier valuing does integrate and combine the performance metrics of a single buyer-supplier relationship with the PSM strategy and the corporate strategy as well as the purchasing and company value creation process. Supplier valuing ensures that decisions concerning partnering with suppliers are aligned with the target supplier lifetime value to improve PSM and company performance (Arnold *et al.*, 2005).

There exist several problems with this approach which are also mentioned by the authors (Arnold, 2004). Most critical for them are the measurement issues. The measurement of future outpayments and inpayments, the definition of the correct relationship duration or the appropriate discount rate are crucial to calculate an accurate supplier lifetime value. Unfortunately, all variables are extremely difficult to calculate, e.g. different specific discount rates would have to be determined depending on the buyer's risk at stake during the relationship which heavily depends on manifold factors. Another important aspect is the dynamically changing target supplier lifetime value because it is directly connected to a company's strategic priorities. Consequently, it might be necessary to include the supplier's ability to follow the buying firm's strategy changes for different strategy change scenarios with a defined probability to determine a concrete target supplier lifetime value. As a result, the proposed approach is theoretically valuable but difficult to implement in practice. Nevertheless, *Arnold* and *Eßig* highlight the importance of a permanent gap evaluation concerning required supplier capabilities and existing capabilities. Their concept supports the proposed capability-gap analysis as a starting point for a strategy-based supplier selection.

Möller and *Törrönen* (2003) address with their approach the problem of evaluating the value creation potential of strategic suppliers. They discuss the types of value which may be achieved through or with a supplier, identify the criteria that enable or impede value creation, and propose a conceptual framework for the assessment of a supplier's value creation potential. In essence, they argue that a customer could use a supplier's current capability profile as a proxy indicator of how suitable that particular supplier is for specific value creation projects.

According to their approach, a supplier provides value for the buying firm in several ways. In its simplest form, this value is reflected by the market price. When the value creation requires joint efforts, the focus depends on the characteristics of a particular supplier-buyer relationship. They basically classify business relationship functions into direct and indirect functions. Direct functions describe the immediate cost-and-revenue effects of a supplier relationship for the buying firm. Indirect functions are more difficult to measure because their impact is realized through the linking of the supplier-buyer dyad to other players. Therefore, the direct functions may be realized within a specific dyad, whereas the indirect functions rely on the linkages provided by the firms to a larger network environment (Möller *et al.*, 2003).

From a measurement perspective, they propose to analyze the effects of actions within a relationship on four levels. The first level of effects in a relationship refers to activities which can be realized without any, or with only minor adaptations among the relationship partners. The second level comprises the generative effects of a relationship. They show the influence of activities in which adaptation by the actors is a prerequisite. For example, the success of Wal-Mart can be partly ascribed to its systematic use of intensive buyer-supplier relationships in terms of information system standards. By leveraging its negotiation potential, Wal-Mart forces suppliers to adapt to its efficient demand-pull system achieving highly efficient processes. The third level concerns the effects on the relationship portfolio of the supplier and/or buyer. For example, by becoming engaged in a major co-operational venture with a specific supplier, a buying firm may destroy its possibilities to develop relationships with new customers which compete with the preceding specific supplier. On the other hand, the co-operational venture may also have a positive reference effect on the new customers that are not competing with the co-operational partner. Finally, the fourth level illustrates the effects of actions within a relationship on the wider network of the supplier and/or buyer. If a buyer develops new technology with a particular supplier, a level 2 activity, this may become a new industry standard and be adopted throughout the network, thus providing positive returns for the initiator (Möller *et al.*, 2003).

As a result, the complexity involved in determining any major changes in a supplier-buyer relationship is very high due to the number of contingencies and the relatively long time horizon influencing the realization of the benefits and costs of the activity. From a practicability perspective, *Möller* and *Törrönen* (2003) propose the use of approximations in suppliers' value evaluation.

They classify a supplier's value in the three dimensions efficiency, effectiveness and network. Efficiency refers to the efficacious use of current resources, in other words, getting more out of the applied resources. A supplier consistently offering a better price for a standard item operates more efficiently than its competitors, and it may also have a better supplier portfolio itself. Efficiency value may be present in a supplier-buyer relationship at the first level, respectively without any adaptations. By adjusting their operations (level 2) in order to achieve a better match between their processes, companies often realize considerable efficiency gains as proven by just-in-time applications in numerous industries (Möller *et al.*, 2003).

Effectiveness refers to a company's ability to invent and produce solutions that provide more value to customers than existing offers. A single supplier may produce new effective solutions, although this is becoming rare due to the difficulties and costs involved in mastering the multiple technologies generally necessary. Nowadays, the development of products and processes commonly takes place through cooperation between the supplier and the buying firm in cross-functional teams. This kind of project requires mutual adaptations (level 2) which may affect the current supplier/buyer portfolio (level 3), or even the larger network (level 4) (Möller *et al.*, 2003).

The last basic supplier value is called network and consists of the resource-access function, the scout function, and the market signaling function. The resource-access function describes the network connections of a specific supplier, including its linkages to next-level suppliers, research institutes, and other customers. These linkages may provide the buying firm with resources for enhancing its business processes. The scout function refers to markets and other information that can be obtained from the working environment through a particular supplier. The market-signaling function describes the use of a supplier as a positive reference or signaling effect for the wider network (Möller *et al.*, 2003).

Möller and *Törrönen's* approach suggests a capability-based supplier selection based on their value creation potential. In this doctoral thesis, suppliers' value creation potential is defined by the capability to generate competitive advantages for the buying firm and goes in a similar direction when separating efficiency and effectiveness issues. Furthermore, they propose a very broad set of supplier selection criteria which is directly focused on the creation of value. This aspect will also flow into the development of a strategy-based set of supplier selection criteria. Especially the network functions will also be applied and empirically tested although in different selection criteria factors.

The different described approaches support the assumption that the selection of strategic suppliers is a complex issue. According to the presented concepts, both qualitative and quantitative criteria have to be used by the buying firm in order to reflect suppliers' different characteristics, the capabilities of their respective suppliers, and possibly other related companies. With such an approach, people with different expertise need to be involved. As a consequence, the evaluation of a supplier's (potential) performance should not be seen as a single activity that provides a clear-cut performance grade. Supplier selection rather involves many activities that represent different perspectives leading to a multifaceted outcome. By varying the perspectives in this way, the buying firm might be able to capture all necessary aspects of the focal supplier and existing potentials for joint performance enhancements. The implementation of appropriate measures to ripe this potential requires fine-tuned balancing. For instance, too much emphasis on price capabilities might hamper the development of suppliers' product and process quality or innovation capabilities (Fredriksson *et al.*, 2003).

One of the most important insights for companies is that various selection procedures provide complementary and/or overlapping perspectives on (potential) supplier performance. Complementary perspectives assess the same performance dimension by using different criteria. For example, the performance dimension delivery precision can be evaluated by using two different criteria such as supplier on time at loading dock and number of production plan restrictions set by the supplier.

Overlapping perspectives are at hand when the same criterion is used for the evaluation of different performance dimensions. For instance, the characteristics of the suppliers' internal processes such as equipment capability might be used for assessing both quality and delivery performance.

The assessment procedures for the selection of strategic suppliers provide outcomes concerning their (potential) performance in various dimensions. The assessment results from complementary perspectives might be contradictory. For example, a supplier performing well concerning short-term quality levels might rely on quick-fix-solutions, thus neglecting long-term improvement and investments in quality management. In a similar way, heavy focus on a supplier's transportation costs might lead to sub-optimization of the costs in the overall logistics network.

The various departments of the buying firm emphasize different criteria and do their best to encourage supplier improvements in the area of their respective responsibilities. Therefore, the existence of complementarities among assessment perspectives might lead to a situation where the ambitions of one department of the buying firm to improve supplier performance in a particular capability might reduce performance in other capabilities of the same overall selection dimension.

Overlapping criteria might also cause problems because one criterion is used for the evaluation of two different performance dimensions. Modifications favorable for one criterion dimension might cause then performance losses in the other. Consequently, requirements based on one department's evaluation might erode performance for another department within the buying firm.

To sum it up, assessing a supplier's performance involves various perspectives that lead to complex results and require different skills and resources (Fredriksson *et al.*, 2003). Therefore, PSM managers need to be process owners for supplier selection projects ensuring that the decisions are made for the sake of the overall company performance. For this reason, PSM managers need to apply a broad supplier selection approach. The traditional selection criteria such as price, quality, delivery or service have got into suspicion whether they comprehensively include all relevant success factors in the interaction with suppliers. It seems hard to select the best suppliers in terms of flexibility and innovation through the criteria price, quality or delivery. For today's PSM managers, the wide range of suppliers' capabilities should become the focus; not the delivered products or services because the product and service requirements might change fast compared to most strategic supplier relationships (Lemke *et al.*, 2000). PSM managers need to start with the strategic priorities of the company, identify the required capabilities from the supply base and integrate these requirements into the selection process for strategic suppliers.

3.3 Conceptual framework for a strategy-based supplier selection

The following conceptual framework explains in graphical and narrative form the key constructs, relationships and indicators of the developed research model for the empirical study presented in chapter 4. Based on the conceptual developments and theoretical insights in the previous chapters, a research model has been developed which tests the influence of competitive strategies and supply chain strategies on the importance on a broad range of selection criteria for strategic suppliers. Furthermore, the influence of these selection criteria on four strategic PSM performance levers and their impact on the creation of competitive advantages is tested to support the theory developments made in chapter 2.2. In the following, the hypotheses model is presented in chapter 3.3.1. Each construct represented in the hypotheses model is defined in the following two chapters before discussing the results of the paper-based survey in chapter 4. In chapter 3.3.2, the reflective strategic priorities and performance constructs are jointly discussed in order to emphasize the common requirements for reflectively defined constructs. In chapter 3.3.3, the formatively constructed supplier selection criteria are finally presented.

3.3.1 Hypotheses model for a strategy-based supplier selection

In the following, the hypotheses model of a strategy-based supplier selection is developed based on the insights from the presented literature and conceptual developments of the previous chapters. After a short introduction, the three main hypotheses are presented which correspond with the research questions indicated in chapter 1.2. Firstly, it is explained why the selection criteria for strategic suppliers are expected to be influenced by companies' strategic priorities. Secondly, it is discussed how the selection of strategic suppliers can influence four different strategic PSM performance levers. Finally, it is explained how PSM is expected to contribute to the generation of competitive advantages with its four strategic performance levers.

In today's business, many of the traditional supplier selection criteria still apply, including criteria such as costs, quality, delivery and flexibility but the further integration of PSM into companies' strategic management processes requires a broader approach to the selection of strategic suppliers as discussed in chapter 2.3. Recently, there has been increasing doubt whether the traditional criteria cost, quality, delivery and flexibility really measure the performance spectrum defining the success of supplier-buyer relationships (Lemke *et al.*, 2000). Empirical results indicate that soft, not easily quantifiable selection criteria, such as a suppliers' strategic commitment to a buyer, innovativeness, or relationship factors have a greater impact on performance than hard, more quantifiable criteria, yet are still considered to be less important (Kannan *et al.*, 2002). In order to exploit suppliers to their full potential, it is apparent that organizations may need to make a number of changes in the way suppliers have traditionally been selected (Culley *et al.*, 1999). *Duffy* (1999) postulates that PSM managers need to become so knowledgeable about their current and potential suppliers that they know as much about them as they do about themselves. Although this requirement seems to be difficult to fulfill, she further proposes that PSM managers must become aware of and consider criteria they might not have been considered in the past in order to adapt their supplier selection process for the future.

The proposed strategy-based supplier selection approach in this thesis is thought to provide a conceptual framework that supports PSM managers applying a wider strategic perspective than currently used. Other supplier selection and evaluation tools such as supplier audits or the analysis of the financial stability (Östring, 2003) are important and complementary elements of a professional supplier management approach.

Figure 31 provides an overview of the main three hypotheses A, B and C which are stated as follows:

Hypothesis A: *Companies' strategic priorities in the form of competitive and supply chain strategies influence the relevance of selection criteria for strategic suppliers.*

Hypothesis B: *The relevance of selection criteria for strategic suppliers influences four strategic PSM performance levers.*

Hypothesis C: *PSM's strategic performance levers have an impact on companies' market and financial performance measures.*

Following *Kaufmann* (2004b), the presented approach of this doctoral thesis focuses on the selection criteria for strategic suppliers because it is argued that a one-size-fits-all approach is not supportive for PSM and company performance (Procurement Strategy Council, 2002). Therefore, the developed concept of strategy-based supplier selection is only applicable to suppliers which have a reasonable impact on strategic PSM performance levers and company performance as discussed in chapter 3.2.1.

Figure 31: Research model with three main hypotheses

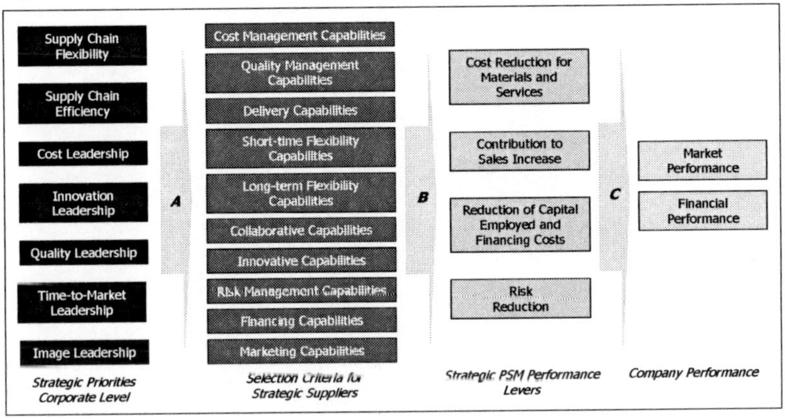

Whether this impact relies on suppliers' specific capabilities in cost reduction, innovation development or risk minimization depends on companies' strategic priorities on corporate or functional level. Therefore, the first main hypothesis is stated as follows:

Hypothesis A: *Companies' strategic priorities in the form of competitive and supply chain strategies influence the relevance of selection criteria for strategic suppliers.*

The overall objective of the supplier selection process is to maximize strategic suppliers' contribution to the generation of competitive advantages. Suppliers differ widely in their capabilities and in what benefits they can bring to a customer. Some suppliers can provide

benefits in terms of cost rationalization whilst others can act as sources of new ideas or improvement practices (Fröhlich-Glantschnig, 1997). It has been shown in previous chapters that PSM and the selection of suppliers strongly influence the implementation of strategic priorities on corporate and functional level. PSM managers should strategically analyze their suppliers to determine the extent to which suppliers' capabilities can contribute to the implementation of their various strategic priorities (Procurement Strategy Council, 2002).
It seems important to highlight that the strategic priorities should not be regarded as either-or trade-offs but rather as dimensions between which a balance should be achieved, determined by the companies' target markets (Lamming *et al.*, 2000). *Christopher* (1992) argues similarly in this point when asking PSM managers not merely to seek to be cost efficient or fast or flexible but to achieve all of these at the same time although not to the same extent.

Figure 32 shows the proposed break-down process from companies' strategic priorities to supplier selection criteria. Starting at the corporate level, a company's strategic priorities, e.g., concerning the position towards competitors, product-/market objectives or supply chain priorities, influence PSM strategy and other functional strategies. The strategic priorities from other functions such as production, R&D, marketing or finance have also to be taken into account as far as they don't form a sharp contrast to corporate level priorities. [28]

Figure 32: Break-down process of strategic priorities

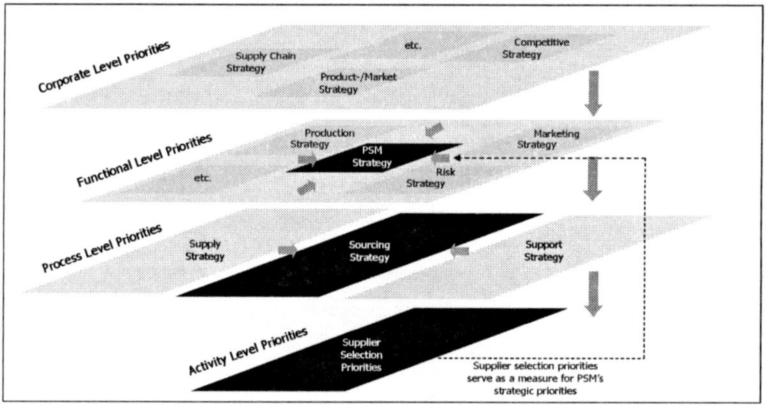

This break-down process model is based on the PSM strategy concept developed in chapter 2.3.2 and the alignment within the strategic management process model as proposed by

28 Although other strategies on corporate level such as product-/market strategies and other functional strategies influence the relevance of selection criteria for strategic suppliers it has been empirically evaluated that competitive strategies and supply chain strategies are the most influential corporate-level priorities (Jahns *et al.*, 2006).

Monczka, *Trent* and *Handfield* (2002) and *Pagell* and *Krause* (2002) in chapter 3. Furthermore, the process perspective taken on the third level is based on the St. Gallen management model adapted for PSM in chapter 2.2.3 and the sourcing definition developed in chapter 2.1.1.

For the selection of strategic suppliers, the sub element sourcing strategy integrates the various strategic priorities from corporate and functional level and determines which priorities are the most relevant for the specific capabilities to be sourced. Most often, the relevance of the strategic priorities will depend on the specific commodity (group) the strategic suppliers need to bring in their capabilities for. PSM managers are responsible for the integration of strategic priorities and prioritize them for the elements supplier strategy, supply market strategy and internal customer strategy as presented in chapter 2.3.2. *Krause, Pagell* and *Curkovic* (2001) additionally identified supplier selection criteria as the best proxy manifestation of, and surrogate for, PSM's overall strategic priorities. Therefore, supplier selection priorities also serve as a proxy for PSM's strategic priorities in figure 32.

While the main hypothesis A postulates a strong relationship between companies' strategic priorities and the relevance of selection criteria for strategic suppliers based on previous conceptual and empirical papers (e.g. Narasimhan *et al.*, 1998; Watts *et al.*, 1995) the main hypothesis B is focused on the impact of the strategy-based supplier selection priorities on four strategic PSM performance levers. This second hypothesis B is based on the assumption that the capabilities of the strategic suppliers have a significant influence on four competitive levers for PSM performance (for similar hypotheses compare also Guinipero *et al.*, 1993; Kannan *et al.*, 2002; Vonderembse *et al.*, 1999). Therefore, the second main hypothesis (B) is stated as follows:

Hypothesis B: *The relevance of selection criteria for strategic suppliers influences four strategic PSM performance levers.*

From a strategic perspective, PSM is responsible to provide the other functions with the necessary supplier capabilities to successfully implement companies' strategic priorities as described in chapter 2.3.2 and consequently generate competitive advantages as analyzed in chapters 2.2.1 and 2.2.2. The support through the right supplier capabilities enables the various functions to optimize their performance and generate competitive advantages. The generation of competitive advantages is finally reflected in companies' financial and market performance measures (Tan *et al.*, 1998). PSM is hypothesized to contribute to companies' performance through four strategic performance levers focusing on PSM's effectiveness

(Gallery *et al.*, 2002; Monczka *et al.*, 1994) rather than efficiency (van Weele, 1984) (figure 33). The analysis of the different strategy theories in chapter 2.2.1 serves as the theory background for this hypothesis. Based on the presented four strategy theory approaches, it was shown that PSM can contribute to the generation of competitive advantages measured as realized monopoly, Ricardian, Schumpeterian and relational rents. In practice, the different rents cannot be directly measured. Therefore, indicators for the realization of monopoly, Ricardian, Schumpeterian and relational rents need to be developed. On corporate level, this is often measured through companies' financial or market performance (e.g. Kannan *et al.*, 2002). A concept of four strategic PSM performance levers is developed in order to link the results of the PSM process level with the outcome on corporate level similarly to the concept presented in chapter 3.1. Based on conceptual papers (e.g. Gallery *et al.*, 2002) and case study analyses (e.g. Entchelmeier *et al.*, 2006) four strategic PSM performance levers are proposed:

(a) Cost reductions for materials and services

(b) Contribution to sales increase

(c) Reduction of capital employed and financing costs

(d) Risk reduction

Each of these four strategic performance levers can be influenced by PSM and any other function (Gallery *et al.*, 2002). For example, the finance department has a huge impact on the reduction of financing costs while the marketing department can contribute enormously to sales increases. PSM is also expected to contribute to all four competitive levers with varying impact through its strategic supply base management and a strategy-based selection of strategic suppliers. This measurement approach has been chosen to support PSM's integration into strategic management processes. A further integration of PSM requires rather an adaptation of PSM to general management performance measures than the introduction of new specific PSM measures into companies' strategic management processes. Accordingly, figure 33 represents the hypothesized influence of a prioritization of specific supplier selection criteria on the proposed four strategic PSM performance levers. For instance, PSM can put pressure on its strategic suppliers as proposed by the market-based view in chapter 2.2.1 through a strong focus on cost management capabilities. This focus is hypothesized to influence the reduction of costs for materials and services. Similarly, companies focusing on transaction costs as part of the cost management criterion experience cost advantages when buying materials and services (Cousins, 2005). Supported by a capability-based view, PSM's prioritization of innovative capabilities is hypothesized to have a strong positive impact on

PSM' contribution to sales increase. The same implications can be drawn from a relational view when focusing on collaborative capabilities as an enabler for idiosyncratic relationships. Based on the discussion of principal-agent theory in chapter 2.2.2, insights from this theory also support the development of specific supplier selection criteria such as risk-rent sharing agreements which enable the joint creation of competitive advantages. Resource-based view supports the impact of strategy-based resource acquisition. A possible case might be the contractual acquisition of single suppliers with superior market access for specific emerging markets the buying firm is targeting during the next time reducing risks and improving market chances.

Figure 33: Supplier capabilities' influence on strategic PSM performance levers

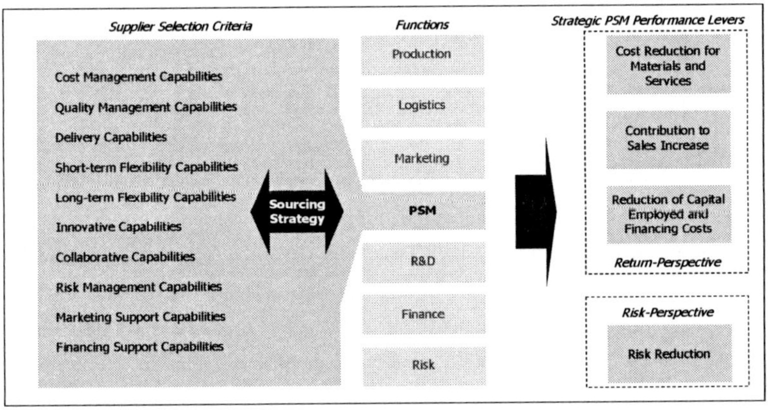

Traditionally, PSM is expected to contribute enormously to cost reductions for materials and services. Recently, a more strategic view acknowledges also PSM's ability to contribute to sales increase and reduction of capital employed and financing costs (Ellram *et al.*, 2002a; Gallery *et al.*, 2002). Finally, the performance picture needs to be completed by a risk perspective (Doyle, 1990). Following *Markowitz'* (1952) value perspective, there is no actual positive impact on a company's overall performance if cost reductions, sales increases or reductions of capital employed and financing costs are only achieved for the sake of an increased risk position.

Consequently, PSM needs to select strategic suppliers which support the improvement of the four performance levers and subsequently the generation of competitive advantages (Weissman, 2004). The reduction of costs for materials and services might be optimally supported by suppliers with excellent cost management capabilities while PSM's contribution to sales increase might be optimally supported by suppliers with excellent innovation capabilities or market entry support capabilities as already hypothesized. In the following, the logic of the applied range of supplier selection criteria is shortly described.

A literature analysis offered an overview of the most applied supplier selection criteria neglecting criteria with only weak relevance for strategy implementation such as a company's financial figures. Cost, quality, delivery and short-term flexibility have been shown to be the most often used supplier selection criteria although the distinction between delivery performance and short-term flexibility performance is not always clear (Stueland, 2004).

In more recent papers, innovation and relationship issues have gained a considerable amount of attention. This development goes hand-in-hand with a more strategic view on PSM and serves as an indicator that the strategic management of the supply base requests more than a focus on quality and delivery as the main strategic priorities of manufacturing and logistics (Pagell, 2004; Scott, 2004). Based on this assumption, an analysis of other strategic company functions such as marketing, finance and risk management showed that further supplier selection criteria might be useful to capture the value of a supplier from a strategic perspective. Marketing support, financing support and risk management capabilities were therefore integrated into the range of selection criteria for strategic suppliers. As a last dimension, the term flexibility has been analyzed in more detail. A literature analysis (e.g. Evans *et al.*, 2001; Gleich, 2001) has provided several distinctive flexibility factors such as delivery flexibility or volume flexibility. Based on the analysis results, the supplier selection criteria short-term flexibility and long-term flexibility have been developed.

Finally, PSM's strategic role can be claimed if it is shown to directly contribute through the four proposed strategic performance levers to the generation of competitive advantages reflected in companies' performance. The third main hypothesis (C) is therefore concerned with the direct influence of PSM's strategic performance levers on company performance. The third main hypothesis (C) is therefore stated as follows:

Hypothesis C: *PSM's strategic performance levers have an impact on companies' market and financial performance measures.*

In chapter 2.3, it has been proposed that PSM needs to be able to contribute to the creation of competitive advantages in order to be accepted from its peers such as marketing, production or finance. As presented in chapter 2.2.1, strategy theory proposes that PSM can contribute to the generation of competitive advantages in the form of four different rents; i.e. monopoly rents, Ricardian rents, Schumpeterian rents and relational rents. Their individual value is difficult to determine in practice but overall rent generation is reflected in a company's financial and market performance. Figure 34 provides a conceptual overview of PSM's influence of the generation of competitive advantages.

Figure 34: PSM's influence on competitive advantages

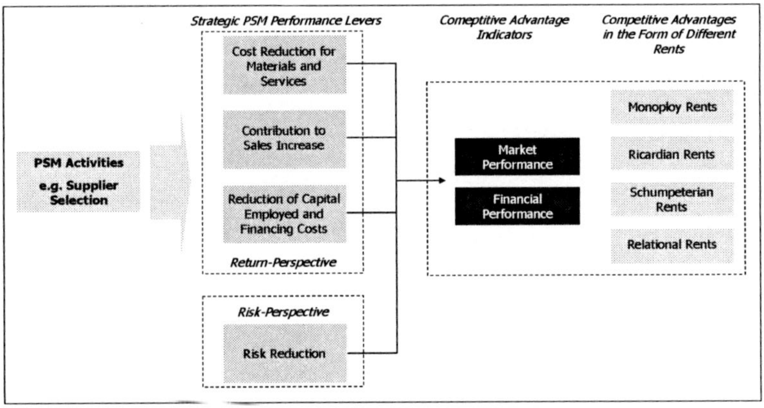

Based on these three main hypotheses, a research model has been developed which integrates the presented hypotheses on a more detailed level. A structural equation modeling (SEM) approach has been chosen as it provides the opportunity to gain a more integrated view on the relationships between the presented elements than with first-generation statistical analyses such as factor or regression analyses (Gefen *et al.*, 2005; Ringle, 2004). For this reason, constructs for strategic priorities on corporate level, supplier selection criteria, strategic PSM performance levers and company performance measures have been developed. Figure 35 provides a graphical representation of the developed research model and single hypotheses. According to *Chin* (1998a), a practice which should be avoided is explicitly providing hypotheses statements for each structural path in complex research models. While each path in the structural model should be explained in the text and be theoretically grounded, the act of stating a null and/or alternative hypothesis for each path is not only redundant and wasteful but can also be confusing to the reader. SEM can easily be described through graphical representation and simple language.

Figure 35: Hypotheses model

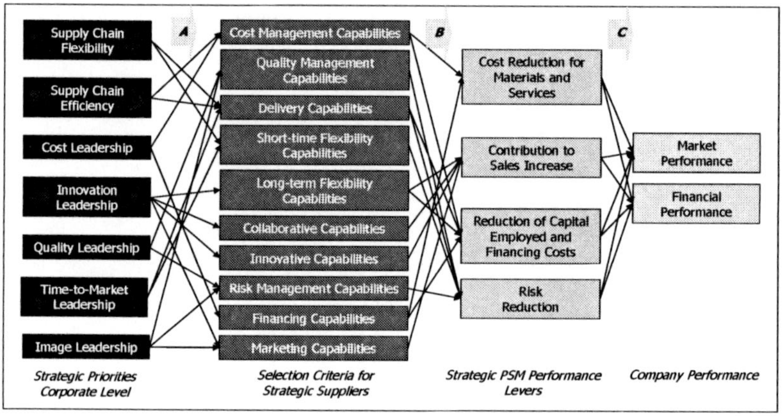

In the following, the three presented main hypotheses (A, B, C) are discussed in detail. Firstly, the proposed influences of competitive priorities and supply chain priorities on the selection criteria for strategic suppliers are explained. Secondly, the impact of strategy-driven supplier selection criteria on four strategic PSM performance levers is discussed. Finally, the influence of the four strategic PSM performance levers on the generation of competitive advantages is analyzed.

In chapter 3.1 of this doctoral thesis, competitive strategies and supply chain strategies were theoretically analyzed for their impact on PSM strategy. Accordingly, each hypothesis is based on these conceptual developments. Firstly, companies with a strong focus on supply chain flexibility are hypothesized to influence the selection of strategic suppliers with strong capabilities in delivery and short-term flexibility because it needs strong logistical support as well as suppliers which can flexibly react on market developments. Companies with a strong focus on supply chain efficiency might prefer suppliers which can support them with an excellent performance in delivery and cost management. Efficiency-focused companies mainly need partners supporting their strategy with logistics capabilities and efficient process cost management (Lamming *et al.*, 2000).

Cost leaders are hypothesized to prefer suppliers with excellent capabilities in cost management and financing support in order to optimally implement their competitive strategy. Innovation leaders might need especially the support of suppliers that are able to build up close relationships, offer long-term flexibility, and provide marketing support and innovation capabilities. These supplier capabilities are supposed to support the buying firm's position as innovation leaders towards their customers. Companies pursuing a position as quality leaders are hypothesized to select strategic suppliers primarily based on their quality and risk management performance because high-level products or services complemented with an appropriate risk management strengthen their image in the customer base. Time-to-market leaders might rely on suppliers with excellent capabilities in delivery and short-time flexibility because they need partners who can quickly react on market developments and have the logistical power to bring the goods to the customers fast. Finally, image leaders provide customers with the opportunity to express a specific lifestyle or message. In order to optimally implement their competitive strategy, they are hypothesized to select preferably suppliers with strong performance in marketing support and risk management.

In the following section, the second main hypothesis is analyzed in detail. For each PSM performance lever, the supplier selection criteria are described which are hypothesized to have an impact. The traditional strategic PSM performance lever cost reduction for materials and services is positively influenced by the selection of strategic suppliers with excellent capabilities in cost management and financing support. While cost management capabilities are a traditional supplier selection criterion financing support seems to be neglected from PSM practice so far. Secondly, the PSM performance lever contribution to sales increase might be supported by suppliers with strong performance in innovation and relationship management, long-term flexibility and marketing support. Innovative suppliers provide the buying firm with new ideas while suppliers, which are able to build up collaborative relationships, support the joint development of new products or process improvements. Suppliers with long-term flexibility are able to support buying firms when entering new geographical markets or customer markets. Finally, suppliers' marketing support is needed to increase sales and market share.

The third strategic PSM performance lever, reduction of capital employed and financing costs, is especially influenced by the selection of strategic suppliers with cost management, long-term flexibility and financing support capabilities. Cost-efficient suppliers can support buying firms in minimizing capital employed. The same is true for short-term flexibility focused suppliers. Additionally, suppliers with excellent financing support capabilities can support the reduction of financing costs. Finally, the selection of suppliers with strong performance in quality management, delivery, short-term flexibility and risk management is hypothesized to impact the PSM performance lever risk reduction. All these capabilities either proactively support the prevention of risks or serve as buffers in case of realized risks.

In this third section, the influence of the proposed four strategic performance levers on the generation of competitive advantages is presented. Competitive advantages are measured in the form of market performance and financial performance (Kannan *et al.*, 2002). Firstly, the PSM performance lever cost reduction of materials and services is hypothesized to generate competitive advantages through its possibilities to directly increase profits benefiting financial performance or to allow price reductions to gain market share. The performance lever contribution to sales increase might impact financial performance through better sales figures leading to optimal capacity management and might impact directly market performance by increasing sales and market share. Thirdly, the PSM performance lever reduction of capital employed and financing cost is hypothesized to have a positive influence on financial performance and market performance through decreased financing costs allowing price reductions or increased margins. Finally, the performance lever risk reduction might improve the financial performance through lower costs caused by less negatively perceived supply chain events and improve market performance through securing the quality of the supply chain (Harland *et al.*, 2003; Kilgore, 2004).

Having analyzed the hypotheses model in detail, a closer look concerning the definition of the proposed constructs seems necessary. Therefore, the reflective strategy and performance constructs are discussed in the following chapter. The formative supplier selection criteria constructs are finally presented in chapter 3.3.3.

3.3.2 Strategy and performance constructs

All constructs applied in this doctoral thesis are based on a five-point Likert scale in the measurement model. Following *Zahra* and *Covin* (1993), the scales used to measure the supply chain and competitive strategy constructs are all anchored in reference to the respondents' key competitors. This approach has two advantages. The first advantage is based on the fact that competitive strategies are comparative terms. By comparing a company's emphasis on a particular competitive strategy relative to that of its key competitors, an accurate assessment of a company's competitive priorities can be applied. The second advantage of this relative measurement approach is the fact that it controls for potentially confounding industry effects (Zahra *et al.*, 1993). In a second step, strategic PSM performance lever and company performance constructs are presented. They are also anchored in reference to the respondents' key competitors therefore excluding interfering industry effects.

In the following, the applied reflective constructs are described. The used measurement items are mentioned but not described in detail for the different strategy constructs because the conceptualization of the constructs heavily relied on measurement items from existing literature:

(1) Competitive strategies

(2) Supply chain strategies

(3) Strategic PSM performance levers

(4) Company performance

Ad (1) Competitive strategies

Research on competitive strategies is well developed. There are many studies providing well established constructs for various competitive strategies. *Porter's* (1980) seminal work provided the first distinction for competitive strategies when distinguishing between cost leadership and differentiation. In order to examine the relationship between a company's competitive position and the relevance of its supplier selection criteria a more detailed differentiation is needed. An interesting approach is provided by *Fuchs, Mifflin, Miller* and *Whitney* (2000) who present a conceptual model for a further differentiation of competitive strategies. Based on their conceptual development and existing literature (Dess *et al.*, 1984; e.g. Green *et al.*, 1993), five competitive strategy constructs have been developed: Cost leadership, quality leadership, innovation leadership, image leadership, and time-to-market leadership.

The concept of cost leadership was popularized in *Porter's* (1980) book on generic competitive strategies. According to his opinion, companies can achieve competitive advantages either through the lowest cost structure in the industry without neglecting other competitive priorities or through the focus on differentiation without neglecting cost structures. Companies following a cost leadership strategy might be characterized by a focus on low production costs, the use of scale effects, efficient business processes and the further development of existing products or services (Green *et al.*, 1993; Zahra *et al.*, 1993).

The main objective of quality leaders is the provision of products and services with permanently high quality (Fuchs *et al.*, 2000). This requirement leads to a different focus in strategic priorities especially influencing companies' production routines. Companies following a quality leadership strategy might be characterized by a focus on zero-defects production processes, high quality standards, high investments in employee training and a broad range of products and services (Dess *et al.*, 1984; Green *et al.*, 1993).

Companies focusing on innovation leadership try to provide path-breaking, state-of-the-art products or services and develop and improve new process technologies (Fuchs *et al.*, 2000). Strategic priorities of pivotal importance for innovation leaders might be a company-wide innovation management, large investments in R&D development and the continuous acquisition of technological know-how (Green *et al.*, 1993; Narver *et al.*, 1990).

Companies vary heavily in their use of marketing, promotion and public relations as competitive priority. Some companies see their customer image as their main tactics to differentiate themselves and their products or services from industry competitors. Creating strong brand identification within a market may have varying utility for companies; e.g. depending upon such factors as the degree to which the companies can generate value through leveraging their visibility across various product lines (Zahra *et al.*, 1993). Therefore, image leaders often offer products or services whose consumption confers social status and signals excellent taste (Fuchs *et al.*, 2000). Companies following an image leadership strategy might be characterized by a focus on the development of strong brands, strong market research and intensive marketing activities (Green *et al.*, 1993; Narver *et al.*, 1990).

Companies focusing on time-to-market leadership define themselves primarily not by the quality or the price of their products but rather by the speed from concept to customer (Ragatz *et al.*, 2002). Especially in industries strongly influenced by fads, trend and other fast cyclic developments, time-to-market leadership might be an important strategic priority (Fine, 1998; Kinder, 2003). Important capabilities of such firms might include the fast identification of market trends, flexible production processes and short delivery times (Heuermann, 2002; Lowson, 2003; Rondinelli *et al.*, 2001).

Ad (2) Supply chain strategies

Supply chain strategy constructs are less developed than competitive strategies. Recently, *Fisher* (1997), *Morash* (2001) and *Lee* (2002) developed and partially empirically tested characteristics for two basic supply chain strategies defined in this doctoral thesis as supply chain efficiency and supply chain flexibility.[29] Their studies and concepts emphasize the importance of the match between the nature of the supply chain and the nature of the sold products and services (Fisher, 1997). In order to support the implementation of a specific

[29] Other researchers work conceptually with the same ideas but with different terms such as leanness and agility (e.g. Towill *et al.*, 2002).

supply chain strategy, the optimal suppliers have to be selected and evaluated differently for each strategic priority. In the following, both supply chain priorities are shortly defined: Supply chain efficiency and supply chain flexibility.

The predictable demand of functional products (Fisher, 1997) implies the need for an efficient supply chain strategy. Companies producing such products might profit from a strong focus on the implied minimization of physical transportation costs. This chosen supply chain priority needs to be supported by the capabilities of the relevant suppliers (Morash, 2001). Based on the above mentioned authors (Fisher, 1997; Lee, 2002; Morash, 2001), the following items have been applied to measure supply chain efficiency: exploitation of scale effects, minimization of transportation and inventory costs, accuracy of sales estimates, broad range of products and services, minimization of transaction costs.

Innovative products (Fisher, 1997) often provide companies with higher margins than functional products but cause severe problems in demand management. Flexibility in the supply chain is a key success factor in order to reap all possible benefits from innovative products (Lee, 2002). Based on the above mentioned authors (Fisher, 1997; Lee, 2002; Morash, 2001), the following items have been used to measure supply chain flexibility: fast information exchange within supply chain, dependency on market trends, length of product and service life cycles, build-to-order concepts, delivery shortages.

Ad (3) Strategic PSM performance levers

Many studies already analyzed the impact of PSM practices on manufacturing performance, supply chain performance or company performance while detailed and direct PSM performance measures have been less developed (e.g. Carr *et al.*, 2002; Das *et al.*, 2000; Goh *et al.*, 1999; Wisner *et al.*, 2000). *Van Weele (1984)* identified four major problems in measuring PSM performance. Firstly, a lack of PSM performance definitions prohibits a general application and comparability in surveys. Secondly, formal objectives and performance standards in PSM are often not clearly defined. Furthermore, PSM performance might be the result of many activities and involved functions which prohibit a clear input-output relationship. Finally, the differences in scope of PSM in many companies differ greatly and partly preclude the development of specific PSM performance measures. Nevertheless, this doctoral thesis tries to directly measure PSM performance. Taking *van Weele's* (1984) identified difficulties into account, a new PSM performance measurement approach is intended to be developed and empirically verified for the first time. Based on *Hughes'* (2005) distinction between relationship, operational, strategic and financial metrics, an approach is chosen which accounts for the research focus of this doctoral thesis and identified measurement problems. Based on his definitions the term strategic PSM performance seems most appropriate for the intended approach. As the PSM function is hypothesized in this study to contribute to the development and the realization of competitive advantages, a multi-dimensional perspective was chosen to determine PSM performance. Additionally, *Gleich* (2001) recommends a strategic perspective for every performance measurement concept. Furthermore, the indicators are formulated in a reflective mode to account for the different responsibility scopes of PSM (Dibbern *et al.*, 2005). This approach enables a linkage between suppliers' capabilities supporting PSM's strategic performance levers and its influence on the generation of competitive advantages represented in companies' financial figures. The focus of the proposed strategic PSM performance measurement lies on effectiveness aspects because efficiency in PSM is shown to be of secondary importance from a rent generation perspective (Jahns, 2006; Monczka *et al.*, 1994).

In many companies, PSM performance measurement is focused on operational or relationship measures. While these metrics are important, few companies relate the operational or relationship perspective to strategic goals and therefore miss many forms of financial value (Hughes, 2005). Furthermore, existing research has often failed to directly relate relationship performance and company's financial performance (e.g. Carr *et al.*, 1999a). The financial impact of PSM goes well beyond cost reductions. It extends to other performance areas such as top-line growth or asset utilization (Ellram *et al.*, 2002a). Other forms of performance measurement such as value-oriented approaches (e.g. Gleich, 2005) might also provide value but they are not in the focus of this doctoral thesis. Research in strategic PSM has not shown to result in many successful studies so far when directly linking PSM activities with company performance (e.g. Ellram *et al.*, 2002b). Starting from a financial perspective on corporate level, the successful generation of competitive advantages (Gälweiler, 1986) is proposed to be reflected in companies' financial performance (Hughes, 2005). Focusing on the PSM level, its contribution to the generation of competitive advantages can be analyzed from a strategic perspective. While companies' financial and market performance can be classified as classical outcome measures PSM's strategic performance perspective might have more a performance lever focus (Ellram *et al.*, 2002a). The proposed strategic perspective by *Hughes* (2005) seems to be the correct performance measurement level for this doctoral thesis because it is directly linked to the financial perspective on corporate level. Moreover, PSM's strategic performance perspective might be conceptualized as a range of performance levers, as proposed by *Ellram* (2002a).

Existing conceptual developments suggest that PSM can directly contribute to the generation of competitive advantages through its impact on cost reduction for materials and services, its contribution to sales increase and the reduction of capital employed and financing costs (Arnold *et al.*, 2005; D'Avanzo *et al.*, 2003; Ellram *et al.*, 2002a; Gallery *et al.*, 2002; Hughes, 2005; Monczka *et al.*, 2000b). According to *Markowitz'* (1952) seminal developments in modern portfolio theory the value of an asset is not only assessed by the expected return but also by the implied risk expressed as standard deviation from the expected mean return. In analogy with his developments, companies might have to assess PSM's contribution not only in terms of overall cost reductions or impact on top-line growth but also in terms of risk reduction towards suppliers, supply markets and the overall supply function (Jahns *et al.*, 2005c; Wen-li *et al.*, 2003). As a result, PSM performance is measured based on the following four constructs which are further described as follows: cost reductions for materials and services, contribution to sales increase, reduction of capital employed and financing costs, and risk reduction.

PSM is often seen as the responsible function to supply the company with the needed goods at lowest possible costs (Pfisterer, 1988). From a broad perspective, cost reductions are not only possible through supplier squeezing but also through the application of more internal-oriented supply techniques. Accordingly, PSM's impact on cost reduction has been conceptualized along four items. Firstly, PSM might reduce costs through clever negotiations with suppliers about prices and discounts. Secondly, cost reductions are possible through efforts to standardize demands from internal customers (e.g. only three laptop models rather than twenty) or to apply design-to-cost or total-cost-of-ownership techniques in order to buy cheaper applications or with less administrative expenses (Hartmann *et al.*, 2005). Thirdly, cost reductions are often possible through company-wide or cross-company demand bundling to increase negotiation power (Meyer, 1990). Finally, specification optimization questions whether all requirements from internal customers such as R&D or production are really necessary or whether simpler specifications might be enough to implement the companies' strategic objectives (Heide *et al.*, 1995).

Beside the classical image of PSM as cost reduction performer, PSM is increasingly seen and required to be able to contribute to top-line growth (Ellram *et al.*, 2002a; Gallery *et al.*, 2002). Accordingly, PSM's impact on a company's sales has been conceptualized as follows: Firstly, PSM might influence sales increase through constantly scanning its current and potential supply markets for already existing supplier innovations in order to improve current products or even develop completely new business solutions (Houghton *et al.*, 2002). Secondly, PSM might foster the development of innovations in cooperation with suppliers leading to superior customer solutions (Carter *et al.*, 2005). Moreover, PSM can reduce development times through a better coordination of the company's own R&D activities and the supplier's R&D activities with techniques such as rapid prototyping or simultaneous engineering (Engel, 2005; Hartmann *et al.*, 2005; Wen-li *et al.*, 2003). In the PC industry, for example, the bottom line impact of a 6 month's product introduction delay can be more than 35% compared to only 5% when overdrawing the development budget by 50%. Especially buying firms competing on customer closeness or time-to-market respectively depend on suppliers which are able to reduce the concept-to-customer cycle times (Sharland *et al.*, 2003). Finally, PSM might use its expertise and knowledge from its discussions with suppliers to improve the company's customer value through innovative ideas or the detection of new material applications or usage trends.

Additionally to cost reductions through the rather physical supply optimization, PSM also has an impact on the level of capital employed and financing costs (Copacino *et al.*, 2004). Most insightful is the impact through the optimization of the inventory. This might happen through improved supply processes such as vendor-managed inventory or simply through the use of market power (Carter *et al.*, 2005; Lee, 2000). For example, suppliers in automotive or electronics industry are often paid 90-120 days after the buying firms' customer has bought the product (Spekman *et al.*, 2004). Furthermore, a professional cash-management through clear agreements concerning maturity dates and an aligned schedule with the company's overall cash flow need supports the minimally required capital (Cavinato, 2000). Thirdly, PSM has influence on the company's overall financing expenses in many ways. For instance, PSM activities are increasingly presented in annual financial statements influencing the analysts' view concerning relevant company risks and therefore influencing the company's credit ranking. Fourthly, PSM managers can contribute to the reduction of capital employed and financing costs through the flexible use of various financing concepts such as leasing instead of buying (Ellram *et al.*, 2002a).

Finally, PSM is hypothesized to contribute to company's overall value generation through its various possibilities to reduce risks for the company. In addition to the first three positively co-notated PSM performance measures, risk reduction is a necessary measurement to correct PSM's impact, e.g., through tough price negotiations or the minimization of development times through simultaneous engineering (Stauffer, 2003). There is no real value added if PSM minimizes purchasing costs by a few per cent but simultaneously increases the probability of supply shortages to a high degree (Jahns *et al.*, 2005c). Based on existing literature (Clouse *et al.*, 2003; Hughes, 2005; Jahns *et al.*, 2005b; Thomson *et al.*, 2000), PSM's impact through risk reduction is conceptualized as follows. Firstly, PSM might reduce supplier-caused risks to an optimal level through periodically conducting supplier risk assessments for strategic suppliers. Moreover, PSM might install early supply risk systems in order to avoid negative impact from occurrences in the supply chain (Kilgore, 2004). Secondly, PSM can reduce internal risks such as bribery, fraud and other legal offences otherwise often leading to a worsening of the company image. Thirdly, PSM has to take care of the currency risks inherent in the supply market. An intelligent sourcing distribution for raw materials or services might prevent the company to loose significant amounts of money due to wrong currency development estimations. Otherwise, risk estimations through value-at-risk concepts might be applied to optimize the born financial purchasing risks (Sanders *et al.*, 2002). Finally, PSM has to ensure an uninterrupted physical supply of the necessary inputs for production and other functions (Hughes, 2005; Piontek, 1999).

Ad (4) Company performance

There are many approaches to measure company performance objectively and subjectively. Most often it is difficult to obtain objective data on financial and operational performance (Injazz *et al.*, 2004). Especially if respondents answer the questionnaire for a large, but specific business unit within a company, the external generation of objective data is almost impossible. Therefore, this doctoral thesis relied on respondents' perceptions of their companies' financial and operational performance. This approach has been shown to correspond closely to objective measures of financial and operational performance (Venkatraman *et al.*, 1986b; William *et al.*, 2003). The buying firms' financial performance has been operationalized through the items return on sales, return on investment, and return on equity as financial key figures (Carr *et al.*, 2002). Additionally, a market-driven perspective has been applied in order to capture the overall company performance. The respective items are defined as sales growth, market share, customer satisfaction and product/service portfolio performance (Carr *et al.*, 1999a; Copacino *et al.*, 2004). Finally, the overall competitive position has been included to capture a single measurement for company performance (Kannan *et al.*, 2002).

3.3.3 Supplier selection criteria constructs

The capabilities of suppliers are diverse. For example, some suppliers which are used to receive design and manufacturing blueprints from their buying firms may lack the capabilities needed to translate the required functionalities into exact product specifications. Other suppliers might have exactly these capabilities but lack the knowledge on how to produce them cheaply (Araujo *et al.*, 1999). PSM managers need to select the best supplier for each task. According to the proposed hypotheses model, different supplier capabilities in the form of supplier selection criteria have to be examined in the study. Many studies (de Boer *et al.*, 2001b; Ellram, 1990; Kannan *et al.*, 2002; Katsikeas *et al.*, 2004; Möller *et al.*, 2003; Swift, 1995) have already integrated various supplier selection criteria into their research concepts. Nevertheless, this study tries to enlarge the range of possible supplier selection criteria in order to more accurately evaluate the possible strategic value of relevant suppliers. In addition to widely accepted supplier selection criteria, rather new selection criteria have been developed:

(1) Cost management

(2) Quality management

(3) Delivery

(4) Short-term flexibility

(5) Long-term flexibility

(6) Innovative capabilities

(7) Collaborative capabilities

(8) Risk management

(9) Marketing support

(10) Financing support

In the following, each supplier selection construct is described with its respective indicators.

Ad (1) Cost management

Cost management is one of the most widely applied supplier selection criteria and has constantly been rated as very important in recent surveys (e.g. Kannan *et al.*, 2002). A possible reason for this constant high rating might be that traditional purchasing thinking is based on the premise that lower prices always add value (Spekman *et al.*, 2004). The cost management construct has been conceptualized as the following supplier capabilities: Competitive pricing, continuous production cost cutting, and continuous process cost cutting.

PSM managers regularly apply some kind of price criterion when assessing suppliers. Although the competitive price might not be the most decisive criterion, especially for strategic suppliers, virtually all supplier selection concepts include a price criterion (Carter *et al.*, 2005).

Continuous production cost cutting is different from competitive pricing as it includes a dynamic component. There is no use in low prices at the beginning of a buyer-supplier relationship if the supplier is not able to significantly reduce the production costs over time. For instance, suppliers' capability to apply joint value analyses or design-to-cost methods can significantly decrease production costs during the product life cycle (Janker, 2004).

Process cost cutting includes the continuous reduction of transaction costs such as administration or coordination costs (Carter *et al.*, 2005). Process cost cutting has been shown to be of high value for the generation of competitive advantages. PSM managers and strategic suppliers have to constantly search for opportunities to reduce process costs of ordering and delivering the respective goods and services (McGinnis *et al.*, 1999).

Ad (2) Quality management

Quality management is permanently rated as the most important supplier selection criterion (Kannan *et al.*, 2002). There exist various approaches to define quality (Harting, 1995). In recent years, the focus of the quality discussion has largely focused on the distinction between an operational quality definition and a customer-defined quality term. Supported by developments in marketing research (Reeves *et al.*, 1994) a customer-oriented quality definition is increasingly used (Buck, 1998). The challenge is the identification of the company's and PSM's relevant customers in order to define the appropriate measurement for the quality criterion. In today's economy, the final producer is seen, from the end consumer perspective, as the ultimate responsible for quality. Every failure is therefore especially the final producer's problem even if a supplier is the source of the failure (Trent *et al.*, 1999). The quality criterion has been conceptualized as the following supplier capabilities: Production within exact specifications, constant specification compliance, and implementation of quality improvement programs.

Production within exact specifications might be classified as a qualifier criterion for strategic suppliers. This measure tends to be common across many supplier selection concepts (Carter *et al.*, 2005; Silveri, 1996). The increased application of ISO (International Standard Organization) certifications in supplier selection shows that the adherence to specifications is an important element (Locke, 1996).

The importance of constant specification compliance can be shown by the following example: A Chrysler supplier changed its ink supplier for the dial it supplied. This new ink affected the electro-mechanical properties of Jeep Cherokee's gauge leading to empty tanks although the display reported adequate levels of gas (Spekman *et al.*, 2004). Acceptable quality levels differ from industry to industry but represent a conditio sine qua non for the selection of strategic suppliers.

While zero defects have become the expected standard in many Western or Japanese companies, production capabilities still widely vary in many companies in the Asian region or in Eastern Europe (Spekman *et al.*, 2004). The selection of strategic suppliers might therefore take into account the implementation of quality improvement programs in order to select the most suitable suppliers.

Ad (3) Delivery

The outsourcing activities of many buying firms cause high dependence on their suppliers' various logistics capabilities including delivery performance (Andersen *et al.*, 2003). Although there still exist manufacturing firms such as automotive companies where logistics performance is not rated as important for PSM managers (Schmitz *et al.*, 2004) many research studies (Handfield *et al.*, 2004; Quayle, 2001) identified logistics delivery performance as a frequently applied supplier selection criterion. Based on existing literature, the delivery construct has been conceptualized as the following supplier capabilities: Delivery on time, delivery sequence compliance, delivery at optimal slot size.

Delivery on time is a key element in today's networked business environment. Buying firms heavily depend on punctual delivery of the goods and services they need to create further value (Disselkamp *et al.*, 2004; Schmitz *et al.*, 2004). The impact of union strikes on production processes in Germany is an example for the relevance of delivery on time performance.

Delivery sequence compliance in the form of just-in-time or just-in-sequence concepts are especially important for production companies such as automotive companies or retailers (Disselkamp *et al.*, 2004). They heavily rely on a continuous and correct flow of supplied products or components because large inventories would be too costly (Boston Consulting Group, 2004).

The delivery of goods at optimal slot sizes is another relevant criterion for the selection of strategic suppliers. For example, optimal slot sizes support buying firms in minimizing their inventory level and improve production processes (Schmitz *et al.*, 2004).

Ad (4) Short-term flexibility

Flexibility is the capability to respond to a changing environment and is increasingly important for the success of many companies (Beamon, 1999). The measurement of flexibility is rather distinctive from other measurement issues due to its highly potential character. Therefore, flexibility actually does not have to be performed by the company in order to exist. Based on *Slack's* (1983) seminal work *Beamon* (1999) developed four types of short-term flexibility: Mix flexibility, volume flexibility, delivery flexibility, and new product flexibility.

Mix flexibility comprises the capability to change the variety of products produced. Generally, mix flexibility is either defined as the range of different product types which may be produced during a particular time period, or the response time between product mix changes (Beamon, 1999).

Volume flexibility comprises the capability to change the output level of products and services produced and refers to the ability to quickly and efficiently adjust output to match demand (Birou *et al.*, 1997; Cuclos *et al.*, 2003).

Delivery flexibility is defined as the capability to change planned delivery dates. This capability allows suppliers to accommodate rush orders and special orders. Delivery flexibility is a very common supplier selection criterion (Schmitz *et al.*, 2004).

New product flexibility contains the capability to introduce and produce new products or modify existing products (Cuclos *et al.*, 2003). It measures the ease with which new products are introduced or existing products are modified by suppliers using time as one of the main determinants. For example, an U.S. company from electronics industry uses an end-customer driven supplier selection system. The rapidly changing nature of the electronics industry means that this company must constantly seek new and innovative ways of maintaining flexibility. A possible way is designing products and sourcing from suppliers by using an approach which considers supply chain order fulfillment for products with extremely short product life cycles. Therefore, two of the important criteria in the company's supplier selection system are ramp-up capability and product specification adaptation (Krause *et al.*, 1999).

Ad (5) Long-term flexibility

As short-term flexibility is necessary to quickly react to immediate changes in operative processes, long-term flexibility is often of pivotal importance for the long-term success of a company (Pearson *et al.*, 1995). It is often stated that the only certain thing for the future is the need to be flexible, especially from a long-term perspective (Duffy, 1999). Long-term flexibility therefore takes a more strategic view. In coherence with the concept of short-term flexibility, a similar approach was chosen for the definition of long-term flexibility: Mix flexibility, volume flexibility, delivery flexibility, and new product flexibility.

Long-term mix flexibility comprises the width and depth of the supplier's product and service portfolio. In a fast changing environment buying firms are increasingly relying on suppliers which are able to adapt their product/service portfolio over time to their changing needs (Rozemeijer, 2000).

Long-term volume flexibility is defined as suppliers' capability to adapt their supplied production volume to the buying firm's product life cycle. Most product life cycle concepts are organized in a five-stage model: design, introduction, growth, maturity, and decline (Rink *et al.*, 2003). The design stage comprises all pre-market activities such as idea generation and concept development. Suppliers can have strong influence as innovation sources and through early supplier involvement as development partners of the buying firms' R&D departments. The suppliers are also encouraged to independently develop new technologies or services. During the introduction stage the suppliers have to work closely with the customer to resolve material defects and modify the product or service based on feedbacks from the customer's market. The growth stage basically shows whether the supplier is prepared for the expected growth rate. Therefore the main question is whether the supplier has the capacities to optimally react when the expected growth is realized or even exceeded. Suppliers should know and be prepared for the expected growth the buying firm is expecting due to their marketing concept and efforts. Naturally, the supply managers have to get these numbers from the company's marketing department and select as well as inform the suppliers accordingly. During the maturity stage, the supplier should be able to support the buying firm either through possibilities for product differentiation or through possibilities for competitive cost reductions. Therefore, depending on the buying firm's strategic objectives the supplier can apply new product innovations, value analysis, process innovations or propose the employment of new materials or components. If sales figures decrease the decline stage starts. During this phase, the pressure on the buying firm and its suppliers increases. Suppliers might be able to slowly reduce the supplied volume and solely focus on cost reductions to keep margins as high as possible. As already mentioned, the stages of the product life cycle are not necessarily consecutive. This means that suppliers might also be able to support the buying firm to stop the sales decline and increase the growth rate again (Rink *et al.*, 2003).

Long-term delivery flexibility is focused on the supplier's capability to organize its supplies for each new production facility of the buying firm. This includes that suppliers are increasingly asked to follow their customers' production facilities in emerging and transition economies such as Eastern Europe, India or China (Monczka *et al.*, 2000b). For instance, some of Whirlpool's largest domestic suppliers had to follow the company into China and India (Laseter, 1998).

Long-term new product flexibility is conceptualized as the supplier's capability to adapt its R&D strategy as much as necessary to the buying firm's strategic R&D requirements (Olsen *et al.*, 1997).

Ad (6) Innovative capabilities

An increasingly important selection criterion for strategic suppliers is their innovative capabilities (Buck, 1998). The identification and integration of innovative suppliers is one of PSM managers' primary tasks (Arnold, 1997). Especially industries with short product development and life cycles such as consumer electronics or even automotive have a high degree of externally sourced R&D services (Bozarth *et al.*, 1998a; Fuhry *et al.*, 2003). In such industries, the need to quickly acquire components and materials regardless of the location is getting critical in order to survive. It seems important to highlight that not only suppliers which are supplying modules or even complete systems have a major influence on their customers' competitiveness (Hartmann *et al.*, 2001). Companies are also increasingly sourcing innovations through the detailed assignment of suppliers' R&D departments or the purchase of existing R&D output concerning both product and process innovations (Fuhry *et al.*, 2003). The innovative capabilities construct has been conceptualized as the following supplier capabilities: Significant financial investments in research projects, joint development of new products or services, independent development of new products or services, and identification of innovations along the supply chain.

Strategic suppliers might provide value through their capability to significantly invest their own monetary and personnel resources into research projects supporting buying firms' strategic priorities (Carter *et al.*, 2005; Disselkamp *et al.*, 2004; Gassmann, 2003).

Many buying firms assess strategic suppliers' innovative capabilities through their willingness to share and jointly develop new innovations concerning technologies or business solutions (Carter *et al.*, 2005).

The capability of suppliers to independently develop new products and services in order to provide their buying firms with new ideas and business opportunities is increasingly rated as important (Janker, 2004; Katsikeas *et al.*, 2004). For instance, retailers with strong brands and market power such as Boots and J. Sainsbury PLC. shift the pledge of innovation onto their suppliers by not only contracting out their information technology but also the identification of innovative uses of IT in business processes (Linder *et al.*, 2003).

Suppliers' in-depth knowledge of their own supply markets can provide a buying firm with specific technical expertise not available in-house or through its own PSM managers, or with alternative new sourcing solutions (Houghton *et al.*, 2002; Ulaga, 2003). Innovative suppliers are actively searching for new materials or technical solutions improving their supplied products or services for the buying firm.

Ad (7) Collaborative capabilities

Suppliers and buyers are increasingly working together in a collaborative way rather than adversarial negotiating about prices. A collaboration-based relationship can be defined as a mutual, ongoing relationship involving a commitment over an extended time period and a sharing of information, risks and rewards (Ellram, 1990). In order to successfully work together, suppliers might need the following capabilities: Joint objective formulation, information and knowledge exchange, build-up of trust, cultural compatibility on management level, cultural compatibility on shop floor level, and efficient/effective conflict resolution.

Goal congruence and formalized risk-reward agreements are important for many buyers when deciding to go for a collaborative relationship with a supplier. These agreements refer to arrangements where a buyer and a supplier work together and mutually expose themselves to potential losses or profits (Thomson *et al.*, 2000). While objectives might change over time and need to be jointly reformulated, an initial fit between the buyer and the supplier in terms of risk and rent sharing attitude (Ellram, 1990; Jahns *et al.*, 2005c) as well as future plans are seen as a prerequisite to establishing a collaborative, long-term relationship (Ellram *et al.*, 1995b; Harland *et al.*, 2003). The mutual acceptance of formalized risk-reward agreements requires that the buyer and the supplier clearly know each others capabilities that can contribute to the joint success (McGinnis *et al.*, 1998).

Information and knowledge sharing is important when working collaboratively to develop new products and services or improve the physical supply chain flow (Ellram, 1990). Sharing demand forecasts and the exchange of technical information are therefore key elements of the knowledge sharing between buyers and strategic suppliers. Furthermore, suppliers which have an open communication culture might better understand buying firms' needs in terms of products or service requirements (Ellram *et al.*, 1995b).

Trust is not simply an input to a relationship. It is even more an outcome of the different interaction activities between buyers and suppliers (Hines, 1996). Thus, cooperative arrangements lead to successful endeavors building trust but most companies would not undertake these activities which build trust without a sufficient level of trust initially (Johnston *et al.*, 2004). This makes its development so difficult and its measurement so multidimensional (Ellram, 1990). *Whipple* and *Frankel* (2000) suggest two forms of trust: character-based trust and competence-based trust. While character-based trust examines qualitative characteristics of behavior inherent in buyers' and suppliers' strategic philosophies and cultures, competence-based trust involves specific operating behaviors and day-to-day performance. Character-based trust constitutes of five elements (Whipple *et al.*, 2000): integrity as the level of honesty between buyer and supplier; identification of motives as the true intentions of buyer and supplier; consistency of behavior as the reliability and predictability of the buyer's and supplier's action in different situations; openness as the willingness to be honest about problems, and discreteness as the willingness to confidentiality maintain of strategic plans and key information. The four parts of competence-based trust can be determined as follows (Whipple *et al.*, 2000): specific competence as specialized operational knowledge and skills; interpersonal competence as an individual's capability to effectively perform his or her responsibilities and work well with others; competence in business sense as a broad experience base beyond a specific area of expertise, and judgment as decision-making capability.

Cultural fit between the buyer's and supplier's top management is often perceived as critical for collaborative relationships because the top management sets the company's strategic direction and the overall tone of the relationship (Ellram, 1990). Supplier's top management should be adaptive to the buyers' top management cultures in order to successfully foster all important relationships.

In addition to the cultural compatibility on management level, people who work together on an operative level also have an important influence on the collaborative success for both parties (Ellram, 1990). For instance, issues such as the 'not-invented-here' syndrome (Katz *et al.*, 1982) can heavily impact the integration of potential innovations into companies' production processes and foster or hinder the generation of competitive advantages.

Conflicts inevitably occur when two parties have to work together but possess different and potentially contrasting objectives. An efficient and effective conflict resolution is related with the cultural compatibility on management and shop floor level. The supplier's capability to solve problems fast and without negative influence on the collaborative relationship is important both on shop floor and top management level (Patterson, 1996).

Ad (8) Risk management

Current developments such as shortened product and technology life-cycles as well as decreasing degrees of value added are leading to complex and differentiated supply chain structures increasing the strategic vulnerability of many companies (Krapfel *et al.*, 1991; Slack, 1983). One important consequence is that risk issues have become more obvious along the supply chain and managers have started to realize that companies often heavily depend on suppliers to achieve their risk management objectives (Christopher *et al.*, 2004; Spekman *et al.*, 2004; Zsidisin, 2003). In the past, companies produced in-house, sourced locally and sold nationally. With a decreasing degree of value added and the saturation of Western consumer markets, many firms experience an increased exposure to externally controlled risks and the need to integrate their often globally distributed suppliers into their risk management activities (Harland *et al.*, 2003; Jap, 2001). As a result, suppliers are increasingly perceived as a source of risk but they can also support the buying firm in avoiding risks, minimizing the impact of risks or bear realized risks within the supply chain (Kilgore, 2004; Sarkis *et al.*, 2002). From this positive perspective, suppliers' risk management activities are an increasingly important part of their overall performance whereas all possible sources of problems caused by suppliers belong to the negative risk aspects (Jahns *et al.*, 2005b). In the following, supplier selection criteria are presented which measure the positive impact of suppliers on the deployment of a professional supply risk management strategy. The risk management construct is based on risk avoiding, risk minimizing and risk bearing activities as supplier capabilities which allow the transfer of risk management activities from a company to its strategic suppliers (Jahns *et al.*, 2005c, p. 41). Strategic suppliers might have the following capabilities in order to support a company's supply risk management activities: Financial risk avoidance, financial risk minimization, operational risk avoidance, operational risk minimization, and risk bearing (financial stability).

Global events such as the currency devaluations occurring in Asia in 1997-1998 or the SARS epidemic can cause substantial financial risks for buying firms (Guinipero *et al.*, 2004a). Financial risk avoidance can happen, for instance, through suppliers' capabilities to offer buying firms globally dispersed production facilities in order to avoid financial risks due to unexpected currency developments.

Financial risk minimization is focused on the reduction of realized financial risks. Strategic suppliers can support buying firms through the joint development and implementation of countermeasures to compensate for negative financial results due to unpredictable market developments (Carter *et al.*, 1993). This includes the joint determination of best pricing currencies or hedging activities such as forward and future contracts or currency options (Locke *et al.*, 1994). A joint agreement on the distribution of risk minimization activities can improve the risk management efficiency and effectiveness of both companies.

Operational risk avoidance through strategic suppliers focuses on the integration of buffers in each value creation step in order to prevent any kind of negative impact of operational problems on the supply chain. As a consequence, efficiency considerations and risk management activities have to be balanced (Guinipero *et al.*, 2004a). A smart way of operational risk avoidance is the coordinated postponement production with suppliers in order to react flexibly on changes in customer demands. For instance, Benetton practices this concept with considerable success for its T-shirt production (Jüttner, 2003).

Suppliers' operational risk minimization capabilities improve the steady supply of the required goods and services. An example for operational risk minimization with suppliers provides a British company from electronics industry. The company first reduced its number of active cable suppliers from 84 to 4. These four suppliers were then brought together by its PSM managers to discuss how they, as a group, would secure the company's global cable needs. The four suppliers were encouraged to be mutually supportive for each other in case of operational problems. The buying firm fostered contingency planning among the four to ensure supply deliveries in case of fire, flood etc. using the suppliers for the active operational risk minimization for a specific part of the buying firm's supply chain (Krause *et al.*, 1999).

Proactive risk management such as risk avoidance or impact minimization can contribute to the generation of competitive advantages (Bozdogan *et al.*, 1998; Spekman *et al.*, 1998). If these activities are not possible or desired, realized risks have to be beard by one or several involved parties. Strategic suppliers can add value through their risk bearing activities. This logic mainly depends on the premise that it is more efficient from an overall value chain perspective that the larger part of the risk is taken by the less risk-adverse company in return for some risk premium such as a greater share of the overall profit. Therefore, the company with the best capabilities for bearing a specific risk along the supply chain should take it in exchange for a market-based compensation (Okamuro, 2001).

Ad (9) Marketing support

Up to now, top line growth has not been a major focus of PSM managers (Michiels *et al.*, 2001). In traditional purchasing environments, sales increase is only influenced by PSM managers' ability to source the components and services required by manufacturing and marketing. A more proactive approach to top line influence involves the selection of suppliers with the following capabilities (Yuva *et al.*, 2002): Information providing about potential customers and markets, support buying firms' entry into new product markets, reference potential, fulfillment of ethical/social standards, and support buying firms' entry into new geographical markets.

Information provided by suppliers are often critical elements for buying firms (Spekman *et al.*, 2004). Their knowledge about current or potential customers and markets might be valuable for the buying firm in order to improve existing products or develop new services. For example, suppliers might already operate in emerging or transition economies while the buying firm still evaluates the success potentials for their products in these respective markets. In such cases, suppliers can support the buying firms with their market knowledge.

Strategic suppliers can support buying firms in entering new product markets with their specific expertise and therefore support a company's diversification activities. Especially in horizontal or lateral diversification efforts (Ansoff, 1958), failure rates are high and suppliers' experiences with targeted customer or product markets has potentially a high value (Amit *et al.*, 1988). This may also include their ability to avoid mistakes in product development or product introduction.

Rather than building brands on their own, many marketing managers in a large number of industries exploit the possibility of achieving a better brand management by the cooperation with other companies and the usage of their brands. Co-branding is a comprehensive term referring to the cooperation between two or more marketable items that in some way connect representations of several brands in the marketplace. Brands can refer to elements such as company names, product designs, logos etc. Although branding is still mostly related to marketing managers, PSM managers should know about ingredient branding possibilities and requirements. This concept is based on the fact that a supplier can add to the total value of the buying firm's product or service by integrating its brand into the buying firm's product or service (Bengtsson *et al.*, 2004; Weigand, 1997). One of the most famous examples is the Intel Inside program in cooperation with computer manufacturers. The potential benefits for companies can include better access to distribution channels, higher operating margins, faster turnover rates or an improved image of their own products or services. The most important disadvantage for companies is the partial lost of control over their own brand and the forced connected development of the ingredient brand supplier (Bengtsson *et al.*, 2004).

Aspects of growing importance are social, ethical, and environmental issues (Young *et al.*, 2001). Firms are increasingly under pressure to take responsibility for their social, ethical, and environmental impacts on society. Buying firms have to ensure that not only their own operations but also their suppliers fulfill the required standards in a global market place. For this reason, buying firms are starting to recognize that they need to identify, understand, and manage these issues not only within their own operations but also along the overall supply chain. These aspects go far beyond compliance concerns as companies such as Nike, many automotive companies have learnt the hard way. Public exposure to such issues as child labor or environmental damaging practices in the buying firm's own operations or supplier's activities can dramatically affect customer's perception of a company and their loyalty. Especially social and ethical issues are perceived to be less tangible and difficult to measure than environmental concerns (Young *et al.*, 2001). Therefore, many buying firms still focus their sustainability supplier management activities on environmental issues. These often include requirements of standards such as ISO (International Standard Organization) 14001 or EMAS (Environmental Management and Audit Scheme). For example, some companies such as IBM, HP, Lucent or Ericsson expect their suppliers to have programs on many elements of environmental performance and include detailed environmental obligations in their contracts (Young *et al.*, 2001). Nevertheless, in today's business it seems important for companies to assess their suppliers not only on environmental issues but also on ethical and social aspects.

Especially industries with shortened product life cycles require the almost simultaneous introduction of new products and services in all relevant (geographical) markets. Communication technology and global supply chains have ruled out the earlier approaches of introducing a new product or service into a company's home market first and then exporting it as local market saturation starts. Today, successful companies must introduce their products or services in all relevant markets simultaneously in order to win market share and preclude entry by competitors (Bozarth *et al.*, 1998a). Strategic suppliers with global production or distribution networks can optimally support their customers when entering new geographical markets.

In recent years, many companies also expanded their global supply base in order to comply with domestic content laws. The expanding of global operations means that companies need to develop localized suppliers to meet production requirements for these new markets. The development of new sources also occurs when domestic content laws are deregulated. For example, the deregulation of domestic content laws in many countries and greater global competition are fundamental drivers for companies seeking more competitive suppliers worldwide. A good example is the aerospace industry where local content requirements of the airlines are an important part of the sourcing strategy since aerospace companies have to develop local suppliers with a very high quality level before they can even start to bid for new contracts (Krause *et al.*, 1999; Moser *et al.*, 2005).

Ad (10) Financing support

The financial stability of suppliers is a standard (passive) criterion in many selection approaches (e.g. Östring, 2003). Suppliers can also actively impact a buying firms' performance through their support in financial aspects. They can influence the cash requirements, the capital employed as well as analysts' perception of the vulnerability of the supply chain. Therefore, the financing support construct has been conceptualized as the following supplier capabilities: Innovative financing instruments, reduction of capital employed, Sarbanes-Oxley Act compliance, and cash management support.

Especially for big deals, the financing aspects are an important part of the negotiation and decision processes. Concepts such as sale-and-lease-back and other financing instruments can influence companies' cash and tax situation. Strategic suppliers with large volumes might be selected based on the innovative financing instruments they offer their customers (Krüger, 2004).

The level of capital employed is an important factor for many financial performance measures. If PSM managers can support the improvement of the capital employed situation by selecting suppliers with the respective capabilities they can improve the buying firms' financial measures. Beside the influence on the inventory level, suppliers can, for example, support buying firms' through insourcing of activities earlier managed by their customers (Cavinato, 2000).

The Sarbanes-Oxley Act (SOX) demands from public companies traded on one of the U.S. exchanges to conduct business in an ethical and open manner. This also implies that their operations are free from conflicts of interest. Therefore, SOX will cause many buying firms to prohibit staff from accepting even small gifts (Cavinato, 2003). Even more important, managers can no longer take the word of employees and accountants about finances without ensuring that the buying firm's buying transactions are properly managed (Grimes, 2004). As a result, buying firms need the support from their suppliers to clearly document and properly implement compliance procedures to efficiently fulfill the SOX requirements (Harland *et al.*, 2003).

PSM has a direct impact on the amount of cash through its payment terms and agreements with its supply base. Suppliers might provide an excellent source of rate-free capital. This aspect might especially flow into consideration for the selection of strategic suppliers with large supply volumes (Cavinato, 2000).

3.4 Interim results: A research model for a strategy-based selection of suppliers

A strategy-based supplier selection involves manifold activities. Firstly, the implications of strategic priorities on corporate level need to be broken down into detailed activities for PSM. Secondly, the interests of PSM's internal customers need to be integrated. This balance of strategic priorities calls for a framework answering organizational, process and content questions. The focus of this doctoral thesis lies on the content questions. Nevertheless, organizational aspects concerning PSM's responsibility and the involvement of other functions in supplier selection are discussed. It is shown that effectiveness in strategic supplier selection can only be achieved if other functions are involved. This is important for PSM's strategic role as strategy is more concerned with effectiveness than efficiency. The weighting of the supplier selection criteria is another pivotal element of the supplier selection process. Four possible approaches to the selection of strategic suppliers are discussed. Although many other weighting concepts exist, a combination of the weighted point method and the analytical hierarchy network method represent a weighting approach which integrates practical feasibility for PSM managers and the required objectiveness in the break-down process of strategic priorities.

The content aspects in the form of supplier selection criteria represent the main focus of this doctoral thesis. Based on the analysis of the strategic implications from competitive strategies and supply chain strategies as well as the analysis of current supplier selection and other supplier management concepts in the literature, a research model has been developed which is tested in the following chapter.

Five competitive strategies have been identified: cost leadership, quality leadership, innovation leadership, image leadership and time-to-market leadership. Each of these competitive strategies is proposed to have different implications for the selection of strategic suppliers and their capabilities respectively. In the same way, characteristics of the company's value proposition model expressed either as supply chain efficiency or supply chain flexibility are proposed to have influence on the required supplier capabilities. Although only part of the empirical research model, the influence of product-/market strategies and other functional strategies on PSM is analyzed as well. It is shown that they may also considerably influence PSM and supplier selection.

The empirical part in chapter 4 is based on the intensive literature analysis concerning strategic priorities, supplier selection criteria, PSM performance measures and company performance measures. Especially the further development of the range of supplier selection criteria and the development of four strategic PSM performance levers for the test in a large-scale study might influence further PSM research. Their detailed description in the last part of chapter 3 allows the examination of a theoretically well-grounded research model with partial least squares analysis.

4 Research methodology, data analysis and empirical results

The following chapter describes the empirical research approach applied to test the developed hypotheses model and examines the results based on partial least squares analysis. In chapter 4.1 the research methodology including research concept and descriptive statistics is presented. Next, the results of the formative and reflective measurement models are analyzed and discussed in chapter 4.2. Finally, the hypotheses model (structural model) developed in previous chapters is tested in chapter 4.3.

4.1 Research methodology

The following chapter describes the reasons for the application of the partial least squares analysis approach as research methodology, documents the questionnaire design and presents the descriptive statistics results.

4.1.1 Research concept and study design

For decades, management research has been focusing on the evaluation of the manifold cause-effect relationships[30] in the business environment with the help of first- and second-generation statistics techniques. Recently, the discussion about the most appropriate research approach for management studies has lead to an increased attention for other research methodologies than the commonly used covariance-based approach (Eberl, 2004). Covariance-based structural equation modeling (SEM) is a second-generation technique and has received high prominence during recent decades. To many management science researchers, the covariance-based methodology is tautologically synonymous with the term SEM (Chin, 1998b). Although there exist different tools which can be used to perform covariance-based analyses such as AMOS the LISREL program has become the most popular one and consequently the term LISREL is often used as a synonym for covariance-based structural equation modeling (Haenlein *et al.*, 2004).

Structural equation modeling approaches are based on a consistent, theoretically derived hypotheses system which contains hypotheses to explain the relationships between latent and observable variables (measurement model) as well as between different latent variables (structural model) as presented in figure 36 (Ringle, 2004).

30 From a pure scientific view, cause-effect relationships can only be detected through controlled experiments. Empirical data from management practice and statistical methodologies such as SEM can only identify positive or negative relationships but not definitely determine cause and effect relationships which have to be supported by theory (Hair *et al.*, 1998).

Figure 36: Structural equation modeling

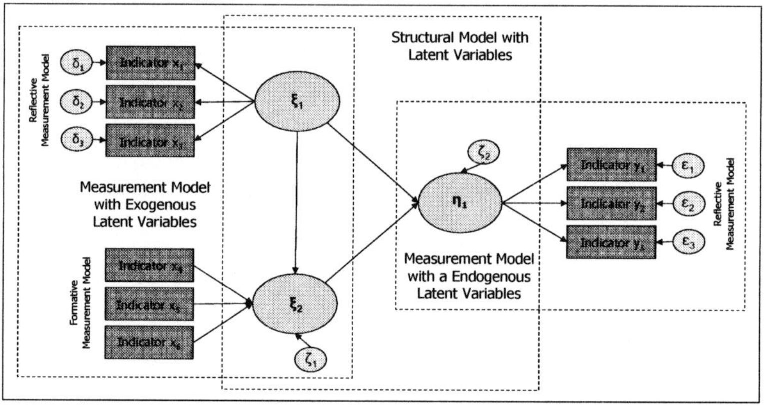

(Source: Adapted from Götz *et al.*, 2004)

Yet, together with the advantages SEM provides comes a higher complexity level concerning its appropriate usage (Chin, 1998a; Homburg *et al.*, 1998). Therefore, a short discussion of the optimal SEM application is provided in the following.

From a statistical point of view, there exist two basically different SEM approaches to estimate the parameters of a structural equation model: the covariance-based (Bentler *et al.*, 1980; Jöreskog, 1970; Jöreskog *et al.*, 1996) and the variance-based methodology (Noonan *et al.*, 1982; Wold, 1975). In this doctoral thesis, the variance-based approach is applied. Therefore, the following explanations will focus on the variance-based approach and explain the reasons for its application.

Variance-based structural equation modeling is almost completely referred to *Wold's* (1975) partial least squares (PLS) analysis which focuses on maximizing the explained variance of the dependent variables through the independent variables. As each structural equation modeling approach, PLS comprises a structural model and a measurement model (figure 36). The structural model reflects the relationship between the latent variables. The measurement model relates the latent variables with their respective indicators. Unlike covariance-based SEM which first estimates its model parameters and then the respective case values, PLS starts by calculating the case values. Accordingly, the latent variables are estimated as linear combinations of their respective indicators and used as proxies (Fornell *et al.*, 1982). The weights applied to determine the case values are estimated so that the resulting case values capture most of the variance of the independent variables explaining the dependent variables

(Chin, 1998b). Accordingly, it is possible to determine a value for each latent variable by calculating a weighted average of its respective indicators. The result is a structural model of latent variables which is approximated by a set of case values and calculated by a set of ordinary least squares (OLS) regressions.

There exist three weighting concepts for partial least squares analyses: centroid, factor and path weighting (Lohmöller, 1989). The centroid-weighting concept is the original procedure used by *Wold* (1975) and is considered to be advantageous when the latent variable correlation matrix is singular. The factor-weighting concept uses the correlation coefficients between the focal latent variable and its neighboring latent variables and maximizes the variance of the principal component of the latent variables when the number of the latent variables goes to infinity. Finally, the path-weighting concept differentially weights neighboring latent variables depending on whether they are antecedents or consequences of the focal latent variable and produces components which both can ideally be predicted and be a good predictor. As a result, the path-weighting concept is often applied with hypothesized causal relations because it is the only procedure explicitly taking into account the directionality of the structural model (Lohmöller, 1989). On the other hand, the factor-weighting concept might be applied if no propositions are made regarding the associations among the latent variables (Chin, 1998b). Finally, it is important to note that the differentiation of the different weighting approaches is not of major importance because it has been proven that the choice between them has only a minor impact on the final measurement model and structural model results (Cassel *et al.*, 1999; Noonan *et al.*, 1982). However, the path-weighting scheme is applied in the proposed research model because it is based on hypothesized directional causal relationships.

After having discussed the basic aspects of PLS, the most important differences to covariance-based approaches are shortly presented and it is explained why PLS represents the appropriate research methodology for this empirical study (figure 37).

Covariance-based approaches use model fitting to compare the covariance structure fit of the researcher's model to a best possible fit covariance structure of the empirical data. The resulting indices and residuals tell how closely the theory model fits the empirical data as opposed to a best-fitting covariance structure. It further integrates detailed assumptions such as multivariate normal distribution about the covariances of the internal residuals and the exogenous latent variables. These distribution requirements are not always existent in the empirical data causing measurement and interpretation problems (Chin, 1998a). Therefore, covariance-based approaches test an a priori specified model against empirical data and emphasize the overall fit of the entire observed covariance matrix with the hypothesized covariance matrix (Fornell *et al.*, 1982; Gefen *et al.*, 2000). In contrast, the PLS concept starts with the empirical data, approximates the latent variables as linear combinations of the indicators and estimates the path coefficients. It is designed to explain variance and is more suited for predictive applications and theory building (Gefen *et al.*, 2000). As a result, covariance-based concepts offers more statistical precision in the context of stringent assumptions while PLS trades parameter efficiency for prediction accuracy (Fornell *et al.*, 1982). Furthermore, PLS has advantages when evaluating single relationships between latent variables and comparing the relative importance of different influence factors (Fassott, 2005).

Beside different prerequisites about the distribution assumptions, research objectives and the sample size, the measurement model has also an influence on the appropriate research approach. There are two basic measurement concepts: reflective (principal factor model) and formative (composite latent variable model) measurement (Chin, 1998a; Jarvis *et al.*, 2003). While the indicators are the results of the latent variable in reflective measurement models, the indicators cause the latent variable in formative measurement models. The latter measurement model is less often used in covariance-based approaches but may also be applied (Chin, 1998b; Ringle, 2004).

Figure 37: Comparison between variance-based and covariance-based analysis

Criteria	**Variance-based analysis**	**Covariance-based analysis**
Objective	Prediction-oriented	Parameter-oriented
Statistical approach	Variance-oriented, e.g. PLS	Covariance-oriented, e.g. LISREL
Assumptions	Predictor specification (nonparametric)	Typically multivariate normal distribution and independent observations (parametric)
Parameter estimates	Consistent as indicators and sample size increases (i.e. consistency at large)	Consistent
Latent variable scores	Explicitly estimated	Indeterminate
Epistemic relationship between a latent variable and its measures	Can be modeled in either formative or reflective mode	Typically only with reflective mode
Implications	Optimal for prediction accuracy	Optimal for parameter accuracy
Model complexity	Large complexity (e.g. 100 constructs and 1000 indicators)	Small to moderate complexity (e.g. less than 100 indicators)
Sample size	Power analysis based on the portion of the model with the largest number of predictors. Minimal recommendation range from 30 to 100 cases	Ideally based on power analysis of specific model – minimal recommendations range from 200 to 800

(Source: Adapted from Chin *et al.*, 1999)

As already indicated, covariance-based approaches require a relatively high sample size to fulfill the multivariate normal distribution (parametric) requirements. PLS applies different resampling methods for goodness-of-fit tests such as jackknifing (Chin, 1998b) or bootstrapping (Zinnbauer *et al.*, 2004) which do not require parametric requirements and therefore less respondents. Furthermore, covariance-based approaches focus on confirmatory and less on exploratory data analysis and require already detailed knowledge about the specific relationships within the structural model (Götz *et al.*, 2004).

The presented study tests different relationships with solid theory background but low empirical knowledge. In addition, there are no detailed assumptions about the distribution character of the data. Furthermore, some of the exogenous latent variables are constructed as formative measurement models. As a result, the PLS research approach has been chosen for the further evaluations of the study.

This decision leads to the question of the required sample size in order to properly evaluate the presented research model. Sample size requirements for PLS research models are usually determined by the following procedure. Firstly, the latent construct with the largest number of formative indicators is identified. Secondly, the dependent variable with the largest number of impacting independent variables is identified. The required sample size can be determined by taking the larger number either of the formative indicators for a construct or the independent variables impacting a dependent variable and multiplying this number with 10 (Chin, 1998b). The number of respondents is not the only factor influencing the quality of parameter estimates. The estimates will approach the true construct scores as the number of indicators per construct and sample size increase. This restricting case is termed *consistency at large* (Wold, 1982). This means that the larger the number of indicators in a construct, the more the essence of the construct is confirmed by the data. But the sample size also needs to increase in order for the covariance matrix to become a better estimate of the population covariance matrix. Therefore, better estimates cannot be obtained simply by increasing the sample size. Furthermore, increasing the number of indicators for a specific construct does not only better approach the true parameter scores but also lowers the standard errors (Chin, 1998b).

4.1.2 Questionnaire design and data collection process

A standardized questionnaire was developed to collect the necessary data for the proposed PLS approach (Stier, 1999). The questionnaire contained only closed and direct questions (Stier, 1999). The applied question types were behavior and characteristics questions in order to focus on current behavior and less on attitudes or opinions (Schnell *et al.*, 1992).

The questionnaire design also includes the determination of the required scale level to evaluate the indicators of the PLS measurement model. Metric scales are optimal for structural equation models but social scientists often do not have the possibility to work with metric scales only. Interval scales such as Likert (Stier, 1999) scales are seen as quasi-metric if they are structured as equidistant (Bortz, 1999; Jaccard *et al.*, 1996). All questions except a few concerning company and respondent characteristics were measured with five point Likert scales. The majority of the supplier selection and evaluation studies used the Likert-type scale (Stueland, 2004). The five answer categories of each Likert scale were described verbally and numerically in order to ensure equidistance (Stier, 1999).

The survey instrument was a paper-based questionnaire which was administered to PSM managers responsible for supplier management. The questionnaire was structured along five sections:

(1) Strategic profile

The first section contained questions concerning the competitive and supply chain priorities of the respondent companies. Furthermore, it was evaluated for how long the current competitive and supply chain strategies were already applied in the respective company.

(2) Capability profile

A first block of indicators was used to evaluate the influence of company-level and functional strategies on PSM strategy as well as the support of strategic suppliers with strategy implementation. A second block of indicators was applied to measure the responding companies' and their strategic suppliers' capabilities concerning image building, quality production, innovation management, operational flexibility and cost management.

(3) Selection behavior profile

The third section evaluated the responding companies' selection behavior towards their strategic suppliers. A first block of indicators measured specific supplier management activities. A second block of indicators examined the importance of 41 selection criteria for strategic suppliers.

(4) Performance profile

In a first block, the performance profile measured the financial and market-oriented company performance. In a second block, PSM's performance was evaluated along four dimensions:

cost reductions for materials and services, contribution to sales increase, reduction of capital employed and financing costs, and reduction of risks.

(5) Company profile

The last section evaluated the industry, the responding person's function within PSM, sales volume, total purchasing volume, percentage of managed spend volume by PSM, percentage of purchasing volume from strategic suppliers and percentage of suppliers classified as strategic.

Prior to data collection, the questionnaire was pre-tested for content validity in two stages. Firstly, two researchers from operations management and four practitioners from strategic PSM were asked to review the questionnaire for ambiguity and appropriateness of the items used to operationalize each construct. Based on their feedback, especially the strategic profile was further developed. Finally, the questionnaire was mailed to and discussed with twelve PSM managers as target group representing different industries and educational levels to achieve a questionnaire with high content validity. The input from the discussions resulted in the rephrasing of some more items while others were deleted completely due to redundancy. The data collection process took place from October 2005 until December 2005.

Strategic PSM literature (Narasimhan *et al.*, 1998; Watts *et al.*, 1995) has not proposed any severe restrictions for the hypothesized influence from strategic priorities on company-level on PSM strategy and supplier selection criteria relevance. Therefore, the presented research model and its corresponding hypotheses have been developed without reference to any industry, specific company size or national constraints (e.g. Kannan *et al.*, 2003; Pearson *et al.*, 1995) in order to confirm the theoretically proposed influence of corporate priorities on PSM strategy and supplier selection priorities (Watts *et al.*, 1995). A further objective of the study was the test of the hypotheses irrespective of company size (in terms of sales volume) as long as at least a souring volume of several millions Euros was achieved. Therefore, a threshold of 10 million euros sales volume was chosen for the potential respondents. Finally, national influences within the German-speaking region were not proposed to have any influences on the hypothesized relationships. Therefore, potential respondents were initially thought to be contacted in Germany, Switzerland and Austria. After the analysis of the potential databases available, only German and Swiss companies were chosen to be included in the sample because only for those countries the needed detailed addresses of the target persons were available to the researcher. The final selection of the addresses was based on a random choice of 440 German and 110 Swiss companies from AZ Bertelsmann Germany and Switzerland with a sales volume threshold of 10 million Euros.

At the end of the data collection process, 112 valid questionnaires were sent back while 32 questionnaires were returned due to incorrect addresses. This equals a response rate of more than 21%.

Non-response bias is a potential limitation of paper-based questionnaire research, even if the response rate is relatively high. A possible test for non-response bias is the comparison of early and late respondents. The latter are more likely to respond to the questions like non-respondents (Narasimhan *et al.*, 1998). A test for non-response bias was conducted in order to ensure that the collected data also represent the non-respondents. The test was conducted by first dividing the sample into two different groups based on the date when the data was received. Each of the sample group consisted of 30 data sets representing those that were received early and late. The difference between these two groups was then tested using a pairwise comparison *t*-test with randomly chosen supplier selection criteria and performance indicators. The test revealed that the two groups of respondents did not differ at a 0.05 significance level. As a result, it was concluded that there was no evidence to suggest the presence of non-response bias in the data.

The next analysis was conducted in response to the fact that a few of the questions were not filled by the respondents. *Little's* (1988) missing completely at random (MCAR) test was used to confirm whether the missing data in every variable appears randomly. This knowledge is important to decide whether missing values have to be treated by simply adopt listwise or pairwise deletion to the missing values or input new values based on specific algorithms. The result was significant at 0.05 levels across all variables. This implied that the missing values in the database were completely random. Therefore, missing values were indicated as -7 and treated as such by the used PLS software through case wise replacement.

4.1.3 Descriptive analysis

As indicated above, 112 companies finally answered the questionnaire. In the following, the characteristics of the responding companies are presented.

The participating companies can be classified according to the following industries (figure 38): the largest segment is represented by the automotive industry with 21%, followed by the mechanical engineering and construction industry (16%) and other services (15%). The food industry includes 7% while the pharmaceutical/chemical industry and the building industry each represent 6%. The electronics industry, aerospace & defense as well as consumer goods comprise 5% each. Less than 5% represented the retail industry (4%), the energy sector (3%) and the financial service industry (2%). Finally, 5% of the responding answers were classified as other industries.

Figure 38: Industry distribution of participating companies

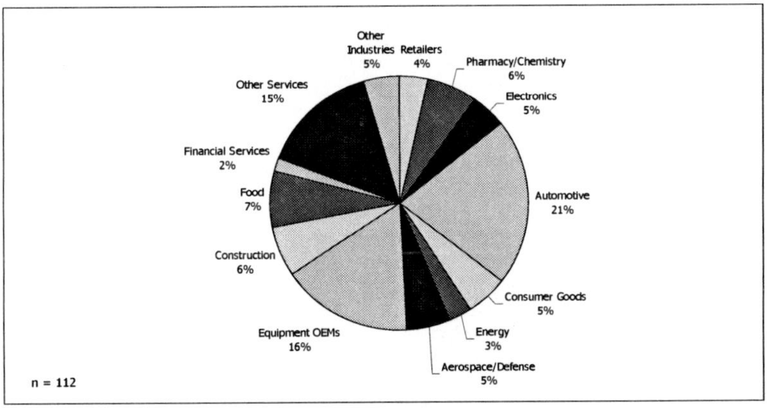

The sample's minimal sales volume was 12 millions Euros while the maximum sales volume was 43 billions Euros. The average sales volume was almost 4.2 billions Euros. This high average sales volume was due to the relatively high proportion of large multinational companies in the responding sample. Nevertheless, companies with a sales volume of less than 100 millions Euros represented 13% of the sample. 35% of the companies had a sales volume between 100 and 500 millions Euros. The companies with a sales volume of more than 500 millions Euros but less than 1 billion Euros represented 25% of the sample. A lower proportion (15%) had companies which had an annual sales volume between 1 and 10 billions Euros. Finally, companies with a sales volume of more than 10 billions Euros accounted for 12% of the sample. Furthermore, the responding companies had on average a degree of value added of 52%.

83 companies from Germany (74%) and 29 (26%) companies from Switzerland answered the questionnaire. This indicates a slightly higher response rate in Switzerland (26%) than in Germany (19%). Based on answers from non-respondents, German companies seem to have more often company guidelines that generally prohibit participation in surveys.

The PSM departments were on average responsible for approximately 69% of the total purchasing volume. This value corresponds to other studies and shows the relatively large impact PSM has on a company's expenses.

The responding PSM managers were hierarchically integrated in the participating companies as follows (figure 39): 35% of the PSM managers were CPOs on corporate level in the responding companies. This is an indicator for the relevance that the selection of strategic suppliers has already in many companies. 20% of the responding PSM managers represented the function of a head of PSM department. CPO on business unit level and professional buyers represented 14% each. Finally, 17% were classified as other function mostly represented by strategic supplier managers.

Figure 39: Hierarchical distribution of participating PSM managers

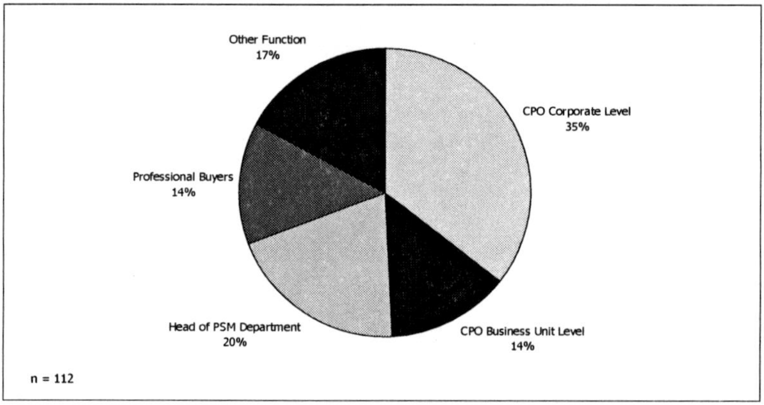

According to the responding PSM managers, the average purchasing volume from strategic suppliers accounts for approximately 60% of the total purchasing volume while about 16% of the supply base is defined as strategic. These values correspond with the general rule of thumb that 20% of the suppliers account for 80% of the total purchasing volume. Whether the purchasing volume of a supplier is a good criterion for its strategic relevance might be another question. The indicated maximum value of 90% and minimum value of 1% for the share of strategic suppliers of total supply base is an indicator for the different definitions of strategic suppliers applied in the companies.

Having analyzed the main characteristics of the sample, the data is tested for normal distribution. First analyses of the distribution histograms indicate a negatively skewed, leptokurtic normal distribution. As a consequence, the Kolmogorov-Smirnov test is applied because looking at histograms is often subjective. The test results indicate that there might be violations of normal distribution characteristics. However, the Kolmogorov-Smirnov test has its limitations because with a sufficient sample size it is likely possible to get significant results from small deviations from normality. Therefore, a significant test doesn't necessarily indicate whether the deviation from normality is enough to bias any statistical procedures (Field, 2005). The further analysis of the actual values of skewness and kurtosis are not informative for this study. Instead, the values need to be converted into z-scores with a mean of 0 and a standard deviation of 1. An analysis of the standardized skewness and kurtosis values indicates significant ($p < 0.05$) values for many of the evaluated indicators.

As the applied research approach in this doctoral thesis is based on partial least squares analysis, the violations of the normal distribution characteristics have no severe impact on the further research procedures. Simulation studies showed that the increase in coefficient biases is only substantial in the case of highly skewed distributions that are not often encountered in realistic applications. Therefore, partial least squares analysis seems to be quite robust to deviations from symmetry (Cassel *et al.*, 1999).

4.2 Measurement model results

The appropriate definition of the measurement model in variance-based and covariance-based approaches is of pivotal importance because the misspecification of the direction of causality between a latent variable and its indicators can lead to inaccurate conclusions about the relationships within the structural model even with reasonable goodness-of-fit statistics (Chin, 1998a; Jarvis *et al.*, 2003). As already indicated, two different measurement models applying multiple indicators of latent variables have been developed so far: formative (composite latent variable model) and reflective (principal factor model) measurement. The choice between the formative and reflective measurement model should be based on thorough considerations concerning the causal relationships between the latent variable and its respective indicators in order to avoid misspecifications (Diamantopoulos *et al.*, 2001).

In the following, the measurement model of the current study is described as well as tested for its goodness-of-fit. While reflective models can be evaluated based on classical test theory applications, formative models have to be evaluated in a different way. Nevertheless, both reflective and formative measurement models are assessed concerning significance levels.

4.2.1 Determination of formative and reflective measurement models

The by far most used measurement model in management research is the reflective model where co-variation among the indicators is caused by the underlying factor (Albers *et al.*, 2006). This approach is based on classical test theory. The causality starts out from the latent variable and changes in this construct are hypothesized to cause changes in all indicators. Therefore, reflective indicators should be internally consistent, uni-dimensional and any two measurement items which are equally reliable need to be interchangeable (Gefen *et al.*, 2000; Jarvis *et al.*, 2003; Rossiter, 2002).

In contrast to the reflective measurement model, the formative measurement model (Curtis *et al.*, 1962) does not assume that a latent variable causes its respective indicators. On the contrary, it is assumed that the indicators have an impact on the latent variable. The causality is defined to flow from the indicators to the latent variable and all indicators together determine the theoretical and empirical meaning of the latent variable defined as a linear sum of a set of measurements. As a consequence, the indicators of a latent variable are not expected to be correlated and can even mutually exclude each other (Diamantopoulos *et al.*, 2001; Jarvis *et al.*, 2003). This implies that dropping an indicator from the formative measurement model usually has a serious impact on the theoretical meaning of the latent variable because this might omit a unique part of the construct.

Jarvis, MacKenzie and *Podsakoff* (2003) have developed an overview with decision rules to determine whether a construct is formative or reflective (figure 40). The first rule relates to the direction of causality between the latent variable and its respective indicators. As already presented, formative indicators are the causes for the latent variable. The second question concerns the interchangeability of the indicators. Formative indicators are not interchangeable because each indicator contributes a specific meaning to the latent variable. The third rule concerns the covariation among a set of indicators. Co-variation is not a requirement for formative indicators but for reflective indicators. The fourth question relates to whether all of the measures are required to have the same antecedents and consequences. Formative indicators do not necessarily capture the same aspects and therefore do not need to have the same antecedents and consequences.

Figure 40: Decision rules for formative and reflective constructs

Question	Formative Model	Reflective Model
1. Direction of causality from construct to measure implied by the conceptual definition - Are the indicators (a) defining characteristics or (b) manifestations of the construct? - Would changes in the indicators cause changes in the construct? - Would changes in the construct cause changes in the indicators?	Direction of causality is from indicators to construct Indicators are defining characteristics of the construct Changes in the indicators should cause changes in the construct Changes in the construct do **not** cause changes in the indicators	Direction of causality is from construct to indicators Indicators are manifestations of the construct Changes in the indicators should **not** cause changes in the construct Changes in the construct should cause changes in the indicators
2. Interchangeability of the indicators - Should the indicators have the same or similar content? - Do the indicators share a common theme?	Indicators need not to be interchangeable Indicators need not to have the same or similar content or share a common theme Dropping an indicator may alter the conceptual domain of the construct	Indicators should be interchangeable Indicators should have the same or similar content or share a common theme Dropping an indicator should not alter the conceptual domain of the construct
3. Covariation among the indicators - Should a change in one of the indicators be associated with changes in the other indicators?	Not necessary for indicators to covary with each other	Indicators are expected to covary with each other
4. Nomological net of the construct indicators - Are the indicators expected to have the same antecedents and consequences?	Nomological net for the indicators may differ Indicators are not required to have the same antecedents and consequences	Nomological net for the indicators should not differ Indicators are required to have the same antecedents and consequences

(Source: Adapted from Jarvis *et al.*, 2003)

These decision rules should support the researcher in deciding whether a latent variable is formatively or reflectively modeled. In addition to these rules, recent articles (Albers *et al.*, 2006; Nicolai, 2004; Varadarajan, 2003) in success factor research have brought up the discussion about the optimal measurement in order to derive useful implications for management practice. Success factor research on management level seeks to identify activities which have a major impact on a wide range of operational and financial performance measures. This implies that managers want to know which specific activities they have to foster or hinder in order to achieve specific targets such as customer closeness or operational excellence. Reflective measurement models do not support this possibility on a detailed level because their indicators have to be interpreted as results of the latent variables and not the other way around. Therefore, formative measurement models, at least for the exogenous latent variables, might be often better suited for success factor analysis (Albers *et al.*, 2006).

In the following, a short overview of the applied constructs and their measurement approach is provided:

(1) Competitive and supply chain strategy constructs (reflective)

(2) Supplier selection criteria constructs (formative)

(3) Performance constructs (reflective)

The measurement models of the constructs are determined based on the described decision rules, prior literature in empirical research and the objectives pursued.

Ad (1) Competitive and supply chain strategy constructs (reflective)

Strategy constructs have traditionally been measured reflectively (Dess *et al.*, 1984; Green *et al.*, 1993; Morash, 2001; Narver *et al.*, 1990). The applied decision rules support this approach because it is rather the strategic direction that causes the relevance of specific company activities such as R&D investments or efficient management of administration processes supporting a specific strategic priority. Furthermore, the applied indicators have been adapted from literature (Green *et al.*, 1993; Narver *et al.*, 1990) that provides a range of interchangeable indicators.

(2) Supplier selection criteria constructs (formative)

Selection criteria for strategic suppliers are applied to select the suppliers which optimally support the chosen corporate strategies. The selection of these suppliers is based on wide range of criteria which are not necessarily highly correlated. For instance, suppliers performing well in the adaptation of their R&D strategy to the buying firm's R&D strategy might not be able to follow the buying firm into new geographical markets. Nevertheless, both criteria are part of the long-term flexibility construct.

Furthermore, each single selection criterion contributes to the respective supplier selection construct in a causal way. Therefore, the direction of causality between the latent variable and its respective indicators is from the single selection criterion to the applied construct. This view is also supported by *Ulaga* who is a strong supporter of a formative measurement approach for supplier selection criteria (Ulaga, 2003). Consequently, from a methodological point of view, a formative measurement approach should be used, rather than reflective measures when modeling supplier selection constructs such as collaborative capabilities or short-term flexibility (Bollen *et al.*, 1991).

(3) Performance constructs (reflective)

Performance is more of an abstract perceptual construct. Therefore, performance might be viewed as reflected in different measures such as market share and relative profitability (Johansson *et al.*, 1994). As a consequence, company performance is reflectively measured from a financial-oriented and a market-oriented perspective. The same considerations are true for the PSM performance constructs.

4.2.2 Reflective measurement model results

The assessment of the goodness-of-fit of a reflective measurement model in a partial least squares approach can be organized along four elements (Götz *et al.*, 2004):

(1) Content validity

(2) Indicator reliability

(3) Construct reliability

(4) Discriminant validity

Ad (1) Content validity

Content validity is defined as the degree to which the indicators of a measurement model belong to the content-semantic sphere of a construct (Venkatraman *et al.*, 1986a). It is usually assessed by explorative factor analysis (Bohrnstedt, 1970). Based on these results, the factors can be further analyzed and interpreted (Götz *et al.*, 2004). Interesting outcomes of an explorative factor analysis are among others total variance explained (TVE) and the Kaiser-Meyer-Olkin (KMO) criterion as first indicators for the appropriateness of a reflective measurement model.

Ad (2) Indicator reliability

Indicator reliability expresses the part of an indicator's variance which is explained by its underlying latent variable. A common criterion is that 50% of an indicator's variance should be explained by its underlying latent variable leading to a required factor loading of 0.7. For relatively new developed constructs even lower factor loadings until 0.4 in PLS models have been accepted (Chin, 1998b; Götz *et al.*, 2004; Hulland, 1999).

Ad (3) Construct reliability

The goodness-of-fit assessment on construct level is even more important than for individual indicators. Construct reliability is examined with the concept of composite reliability.[31]

[31] Other terms for composite reliability are internal consistency, convergence validity or factor reliability.

Fornell and *Larcker* (1981) define composite reliability in reflective measurement models as follows:

$$\text{Composite Reliability (CR)} = \frac{(\sum_i \lambda_{ij})^2}{(\sum_i \lambda_{ij})^2 + \sum_i var(\varepsilon_{ij})}$$

ε_i = measurement error of indicator variable i
λ_i = component loading to indicator variable i
j = run index of all reflective measurement models

Composite reliability can result in figures between 0 and 1. Values larger than 0.7 are commonly accepted but some authors (e.g. Bagozzi *et al.*, 1988) also accept values larger than 0.6 (Krafft *et al.*, 2005; Spanos *et al.*, 2001).

Ad (4) Discriminant validity

An often used assessment criterion for discriminant validity is the average variance extracted (AVE):

$$\text{Average Variance Extracted (AVE)} = \frac{\sum_i \lambda_i^2}{\sum_i \lambda_i^2 + \sum_i var(\varepsilon_i)}$$

ε_i = measurement error of indicator variable i
λ_i = component loading to indicator variable i

According to *Fornell* and *Larcker* (1981), the discriminant validity criterion is fulfilled if a latent variable's AVE is larger than each squared correlation of this latent variable with the other latent variables in the structural model. The cut-off point for average variance extracted is 0.5 (Chin, 1998b; Spanos *et al.*, 2001).

An additional test for discriminant validity is the evaluation of cross-loadings. This implies correlation calculations between latent variable component scores and other indicators besides their own indicators. Discriminant validity can be assumed if all indicators of a specific construct load higher on this construct than on any other (Chin, 1998b).

The relevance of the reflective indicators for their respective constructs should be analyzed based on their loading values (Dibbern *et al.*, 2005). Indicators with low loading values imply that there exist little relationship with the construct in terms of shared variance (Chin, 1998b).

In the following, the applied reflective measurement model constructs and their respective indicators in the PLS structural equation model are examined:

(1) Competitive strategy constructs

(2) Supply chain strategy constructs

(3) Strategic PSM performance lever constructs

(4) Company performance constructs

The cross-loading results for all reflective constructs are presented in the appendix. They support discriminant validity for all constructs.

Ad (1) Competitive strategy constructs

The competitive strategy constructs cost leadership, quality leadership, innovation leadership, image leadership, and time-to-market leadership have been tested for their appropriate measurement (Table 1). Content validity can be confirmed because total variance explained (65.78%) as well as the Kaiser-Meyer-Olkin criterion (0.82) show satisfactory values.

Indicator reliability assessment is based on the constructs' factor loadings. All values are higher than the required cut-off point for established constructs of 0.70. Construct reliability can also be confirmed for all competitive strategy constructs because all values satisfy the CR 0.70 level requirement. Finally, the discriminant validity requirement is also fulfilled with AVE values well above 0.50 for each construct.

Table 1: Measurement model results for competitive strategy constructs

Competitive Strategies ***TVE: 65.78% / KMO: 0.82***				
Construct	**Indicator**	**λ**	**CR**	**AVE**
Cost Leadership	Efficient Administration Processes	0.80***	0.89	0.58
	Efficient Sales Processes	0.76***		
	Low Production Costs	0.78***		
	Use of Scale Effects	0.73***		
	Further Development of Existing Products	0.72***		
	Efficient Capacity Management	0.77***		
Quality Leadership	High Quality Products/Services	0.80***	0.87	0.70
	Zero-Defects Production	0.79***		
	High Quality Standards	0.88***		
Innovation Leadership	Development of Technological Innovations	0.83***	0.92	0.71
	Use of Advanced Technology Knowledge	0.81***		
	Continous Integration of New Technologies	0.79***		
	Investments in R&D	0.86***		
	Company-wide Innovation Management	0.88***		
Image Leadership	Build-up of Strong Brands	0.77***	0.87	0.63
	Focus on Innovative Design	0.81***		
	Intensive Market Research	0.78***		
	Marketing Investment Intensity	0.77***		
Time-to-Market Leadership	Flexible Production Processes	0.74***	0.88	0.66
	Short Delivery Lead Times	0.84***		
	Fast Reaction on Trends	0.87***		
	Fast Trend Identification	0.79***		

* significant at 0.10 ** significant at 0.05 *** significant at 0.01 (two-tailed) ^{n} not significant
TVE: Total Variance Explained KMO: Kaiser-Meyer-Olkin Criterion

Ad (2) Supply chain strategy constructs

Supply chain strategy constructs are less developed in literature than competitive strategy constructs. Therefore, their measurement results aren't expected to be that satisfactory as the previous results from the competitive strategy constructs. Nevertheless, both constructs supply chain efficiency and supply chain flexibility fulfill the content validity requirements with a total variance explained value of more than 53.05% and a KMO value of 0.72 (Table 2). Factor loadings determining indicator reliability are between 0.58 and 0.80 which are satisfactory for relatively new constructs.

Table 2: Measurement model results for supply chain strategy constructs

Supply Chain Strategies ***TVE: 53.05% / KMO: 0.72***				
Construct	**Indicator**	**λ**	**CR**	**AVE**
Supply Chain Efficiency	Inventory Costs	0.80***	0.82	0.61
	Transportation Costs	0.72***		
	Transaction Costs	0.58***		
Supply Chain Flexibility	Build-to-Order Production	0.66**	0.84	0.51
	Short Product Life Cycles	0.73***		
	Fast Information Exchange	0.73***		
	Supply Shortage Costs	0.77***		
	Dependence on Market Trends	0.71***		

* significant at 0.10 ** significant at 0.05 *** significant at 0.01 (two-tailed) ⁿ not significant
TVE: Total Variance Explained KMO: Kaiser-Meyer-Olkin Criterion

Construct reliability can also be confirmed for all supply chain strategy constructs because their values exceed the CR 0.70 cut-off point. Finally, the important discriminant validity requirement is also fulfilled with AVE values above 0.50 for both constructs.

Ad (3) Strategic PSM performance lever constructs

Similar to the supply chain strategy constructs the applied strategic PSM performance lever constructs are based on a rather new measurement concept concerning indicators. Content validity with a total variance explained of 64.93% and a KMO value of 0.77 is fulfilled (Table 3). The indicator reliability assessment shows factor loadings between 0.60 and 0.90 indicating satisfactory values for new constructs. Next, construct reliability requires values above 0.70 which is given for all PSM performance constructs. Finally, discriminant validity is fulfilled with values between 0.58 and 0.67.

Table 3: Measurement model results for strategic PSM performance lever constructs

Strategic PSM Performance Levers *TVE: 64.93% / KMO: 0.77*				
Construct	**Indicator**	**λ**	**CR**	**AVE**
Cost Reduction of Materials and Services	Supply Bundling	0.82***	0,85	0,58
	Price Negotiations	0.77***		
	Specification Management	0.76***		
	Standardization, Design-to-Cost	0.68***		
Contribution to Sales Increase	Reduction of Development Times	0.72***	0,86	0,61
	Innovation Identification in Supply Base	0.90***		
	Innovation Development with Suppliers	0.83***		
	Product/Service Value Improvement	0.60***		
Reduction of Capital Employed and Financing Costs	Cash Management Optimization	0.83***	0,89	0,67
	Financing Optimization of Supply	0.83***		
	Reduction of Financing Costs	0.82***		
	Inventory Optimization	0.76***		
Risk Reduction	Supplier Risks	0.85***	0,87	0,63
	Supply Organization Risks	0.82***		
	Supply Risks	0.82***		
	Currency Risks	0.66***		
* significant at 0.10 ** significant at 0.05 *** significant at 0.01 (two-tailed) ⁿ not significant TVE: Total Variance Explained KMO: Kaiser-Meyer-Olkin Criterion				

Ad (4) Company performance constructs

Financial and market performance constructs are well established in literature and measurement model requirements are therefore high. Content validity is highly satisfactory with a total variance explained of 72.23% and a KMO value of 0.88 (Table 4).

Table 4: Measurement model results for company performance constructs

Company Performance *TVE: 72.23% / KMO: 0.88*				
Construct	**Indicator**	**λ**	**CR**	**AVE**
Financial Performance	Return on Investment	0.91***	0.94	0.83
	Return on Equity	0.89***		
	Return on Sales	0.92***		
Market Performance	Product/Service Portfolio	0.70***	0.89	0.72
	Market Share	0.91***		
	Sales Volume	0.91***		
* significant at 0.10 ** significant at 0.05 *** significant at 0.01 (two-tailed) ⁿ not significant TVE: Total Variance Explained KMO: Kaiser-Meyer-Olkin Criterion				

The indicator reliability assessment shows factor loadings between 0.70 and 0.91 which fulfill the required level. Next, construct reliability values (CR of 0.89 and 0.94) are much above the cut-off point of 0.70. Finally, discriminant validity requirements are also fulfilled with values of 0.83 and 0.72.

In summary, the goodness-of-fit for all reflective constructs can be confirmed.

4.2.3 Formative measurement model results

Guidelines for assessing the goodness-of-fit for latent variables based on formative indicators are seldom. *Diamantopoulos* and *Winklhofer* (2001) have developed a concept containing four relevant issues for successful formative indicator assessment:

(1) Content specification

(2) Indicator specification

(3) Indicator collinearity

(4) External validity

Ad (1) Content specification

Content specification is concerned with the content scope of the latent variable. This specification is of pivotal importance because a formative latent variable is more abstract and ambiguous than a reflective latent variable (Bagozzi, 1994). The definition of the content range of a formative latent variable should consider all facets of the construct (Nunnally *et al.*, 1994). It is to mention that from a practical point of view, an excessive number of indicators is not possible because of the inherent data collection difficulties as well as the increase in the number of parameters when the latent variable is embedded within a large structural model (Diamantopoulos *et al.*, 2001). The procedure to test the appropriateness of the content specification is based on a pre-test with experts (Krafft *et al.*, 2005). *Anderson* and *Gerbing* (1991) have developed two indices to test content specification. The p_{sa}-index is a measurement for the clearness of the indicator assignment to a specific construct while the c_{sv}-index shows the content relevance.

Their calculation formulas are as follows (Krafft *et al.*, 2005):

$$p_{sa}\text{-Index} = \frac{n_c}{N}$$

$$c_{sv}\text{-Index} = \frac{n_c - n_o}{N}$$

n_c = Number of experts assigning the indicator to the correct construct (defined by the researcher)

n_o = Number of experts assigning the indicator to another than the correct construct (defined by the researcher)

N = Total number of asked experts

(2) Indicator specification

For reflective constructs, a set of indicators is randomly chosen from the indicator universe relating to the relevant latent variable (DeVellis, 1991). In contrast, formative latent variables require a census of indicators to cover the entire scope of the respective construct as described under the content specification. The weights of the indicators can be compared in order to assess the individual indicators. Indicators with high weights contribute more to the meaning of the latent variable. Indicators with low weights must not be excluded from the construct because its meaning would change otherwise (Götz *et al.*, 2004). It is important to integrate the significance levels of the indicator weights in order to finally assess their relevance.

(3) Indicator collinearity

Multicollinearity is a problematic issue in formative measurement models because it is based on a multiple regression. Consequently, the stability of the indicator coefficients is affected by the sample size and strength of the indicator intercorrelations. As a result, much collinearity among formative indicators makes it difficult to separate the distinct influence of the individual indicator values on the latent variable. High multicollinearity renders the assessment of indicator validity problematic (Diamantopoulos *et al.*, 2001; Jagpal, 1982). Additionally, if a particular indicator turns out to be almost a perfect linear combination of the other respective formative indicators it is likely to include much redundant information and might be excluded. Multicollinearity can be identified by comparing the correlation matrix of the respective indicators. A multicollinearity identification value can be calculated with the variance inflation factor (VIF). VIF values above 10 are a clear indicator for a problematic multicollinearity situation but researchers can react even at lower VIF values (Götz *et al.*, 2004; Hair *et al.*, 1998).

(4) External validity

A possible way to assess the external validity of a formative measurement model might be the nomological validity concept which has been proposed by *Reinartz*, *Krafft* and *Hoyer* (2004). For this reason, formative constructs might be operationalized in a formative and reflective mode and assessed by a two-construct model (Hauser *et al.*, 1971). Alternatively, the respective latent variable may be linked to other latent variables based on conceptual developments. If the theoretically expected relationship can be proven with the existing data, nomological validity can be assumed (Götz *et al.*, 2004). This doctoral thesis follows the latter approach due to restrictions in the questionnaire.

The interpretation of latent variables measured by formative indicators is based on their respective indicator weights. These weights provide the relative importance of each indicator of its respective construct. The loading values are irrelevant because the intraset correlations for each construct are not taken into account in the further estimation process (Chin, 1998b).

In the following, the applied formative measurement model constructs with their respective indicators for the selection criteria for strategic suppliers are examined and discussed (Table 5).

Content specification for all formative constructs has been evaluated with the support of twelve PSM managers as target group representing different industries and educational levels. The p_{sa}-index and the c_{sv}-index for each formative construct are presented in table 5. The feedback concerning the construction of the latent variables was quite similar to the original set-up as can be inferred from table 5. Therefore, all formative constructs were applied as indicated confirming a correct content specification. Furthermore, external validity for each construct has been analyzed based on its relationships within the nomological network. It will be shown in the structural model evaluation that each construct can be meaningfully integrated into the structural model confirming also external validity for all formative latent variables.

The cost management criterion has originally been measured with three indicators: competitive pricing, process cost reductions, production cost reductions. Due to weak multicollinearity issues (VIP > 3), the indicator production cost reductions was deleted from the cost management construct measurement. Analyzing the indicator specification, it can be concluded that the latent variable cost management is mainly (0.87) determined by the indicator process cost reductions. Competitive pricing has no significant impact (0.18) on this selection criterion for strategic suppliers.

The quality criterion is strongly influenced by the product/service quality (1.17) indicator while the quality specification compliance item had to be removed due to multicollinearity. The capability to implement quality improvement programs has a negative but not significant influence (-0.35) on the quality construct.

The delivery criterion is mostly determined by the indicator delivery time compliance with a weight of. 0.71. The indicator slot size optimization (0.37) has less influence at $p < 0.10$ and the indicator delivery sequence compliance (0.01) has no influence at all.

The short-term flexibility construct has been developed to cover delivery flexibility, production flexibility, product introduction flexibility and mix flexibility. Based on the formative measurement model results, the short-term flexibility selection criterion is mainly determined by the delivery flexibility indicator (1.00). The other three indicators have insignificant weights.

Based on the short-term flexibility definition, the long-term flexibility criterion has been developed. Therefore, it consists of geographical (delivery) flexibility, product portfolio (mix) flexibility, production flexibility and R&D (product introduction) strategy flexibility. Similar to the short-term flexibility construct, geographical (delivery) flexibility (0.43) at $p < 0.10$ but also product portfolio (mix) flexibility (0.44) are the most relevant indicators. Production flexibility has a negative impact (-0.25) and R&D flexibility has a positive (0.42) but insignificant influence.

The innovative capabilities criterion is mostly determined by strategic suppliers' capability to jointly develop new products or services (0.73). Suppliers' capabilities to independently develop new products and services (0.24) and to identify innovations along the supply chain (0.17) have much lower and insignificant weights. Finally, suppliers' capability to invest their own financial and personnel resources into R&D projects for the buying firm has a negative and insignificant influence (-0.13).

Table 5: Measurement model results for supplier selection criteria constructs

Supplier Selection Criteria Constructs						
Construct	**Indicators**	**Weight**	**psa-Index**	**csv-Index**	**VIF**	**NV**
Cost Management	Competitive Prices	0.18^{n}	1.00	1.00	1.00	Yes
	Process Cost Management	0.87***	1.00	1.00		
Quality Management	Product/Service Quality	1.17***	1.00	1.00	1.00	Yes
	Quality Improvement Program	-0.35^{n}	0.92	0.83		
Delivery Capabilities	Delivery Sequence Compliance	0.01^{n}	0.83	0.67	1.87	Yes
	Delivery Time Compliance	0.71***	1.00	1.00		
	Slot Size Optimization	0.37*	1.00	1.00		
Short-term Flexibility	Changes in Current Orders	-0.14^{n}	0.92	0.83	2.54	Yes
	Delivery Flexibility	1.00***	1.00	1.00		
	Short-term Product Introduction	0.09^{n}	0.75	0.50		
	Short-term Production Volume Change	-0.05^{n}	1.00	1.00		
Long-term Flexibility	Geographical Flexibility	0.43*	0.92	0.83	1.56	Yes
	Product Portfolio Flexibility	0.44**	0.83	0.67		
	R&D Strategy Flexibility	0.42^{n}	0.75	0.50		
	PLC adapted Production Flexibility	-0.25^{n}	0.83	0.67		
Innovative Capabilities	Independent Product Development	0.24^{n}	1.00	1.00	2.02	Yes
	Innovation Identification in Supply Chain	0.17^{n}	1.00	1.00		
	Joint Product Development	0.73**	1.00	1.00		
	Resource Investment into R&D	-0.13^{n}	1.00	1.00		
Collaborative Capabilities	Working Culture (Shop-Floor)	-0.57*	1.00	1.00	2.07	Yes
	Joint Target Development	0.20^{n}	1.00	1.00		
	Information and Knowledge Exchange	0.72***	0.92	0.83		
	Fast Conflict Solution	0.47**	1.00	1.00		
	Working Culture (Management)	0.06^{n}	1.00	1.00		
	Trust-based Relationship	-0.13^{n}	1.00	1.00		
Risk Management	Financial Stability	-0.14^{n}	0.83	0.67	2.52	Yes
	Financial Risk Minimization	-0.14^{n}	0.92	0.83		
	Financial Risk Prevention	-0.07^{n}	1.00	1.00		
	Operational Risk Prevention	0.51*	0.92	0.83		
	Operational Risk Minimization	0.67**	1.00	1.00		
Marketing Support	Product Market Entrance Support	0.61**	0.92	0.83	2.11	Yes
	Market Information Provision	0.00^{n}	0.92	0.83		
	Brand Strategy Support	-0.38^{n}	1.00	1.00		
	Geographical Market Entrance Support	0.39^{n}	1.00	1.00		
	Ethical Standards Compliance	-0.00^{n}	0.83	0.67		
Financing Support	Outsourcing Activities Support	0.22^{n}	0.92	0.83	1.58	Yes
	Cash-Management Optimization	0.56***	1.00	1.00		
	Corporate Governance Compliance	0.46**	0.83	0.67		
	Innovative Financing Instruments	-0.08^{n}	0.92	0.83		

* significant at 0.10 ** significant at 0.05 *** significant at 0.01 (two-tailed) ^{n}: not significant
VIF: Variance Inflation Factor NV: Nomological Validity

The collaborative capabilities criterion consists essentially of an effective information and knowledge exchange (0.72) and an efficient conflict resolution management (0.47). Suppliers' capability to adapt their working culture on shop-floor level to the buying firm has a significant negative weight (-0.57) at $p < 0.10$. This results from the fact that different cultures on shop-floor level even a positive impact on the output of a buyer-supplier relationship if conflicts can efficiently be resolved. Both indicators joint target development (0.20) and the adaptation of the working cultures on the management level (0.06) have no significant impact. The build-up of a trust-based relationship has only an insignificant influence (-0.13).

Financial risk management aspects provide only insignificant influence on the risk management construct. This might be due to the fact that financial aspects can be controlled by buying firms themselves without the support of their strategic suppliers. According to the measurement model, this is different for operational risks where buying firms depend more on their strategic suppliers' performance. Important for the risk management criterion are primarily suppliers' capabilities to minimize operational risks (0.67). Their capabilities to prevent operational risks (0.51) are less influential at $p < 0.10$.

Marketing support from strategic suppliers is a rather new supplier selection criterion. The only important and significant item for this construct is strategic suppliers' support in product market entry initiatives (0.61). All other supplier capabilities have insignificant weights. Especially the fulfillment of ethical standards (-0.01) and the provision of market information (0.03) have no impact on this latent variable. Although not significant, suppliers' support of buying firms' brand strategies has a negative impact (-0.38). Finally, the support in geographical market entry initiatives has a positive but insignificant impact (0.39).

The financing support criterion is mainly determined by suppliers' cash-management optimization support (0.56). The second significant capability is suppliers' support in implementing corporate governance requirements (0.46). The support in outsourcing activities has an insignificant weight of 0.22. The provision of innovative financing instruments has a weight of -0.08 which is not significant.

4.3 Structural model results

Having examined the measurement models for the applied reflective and formative constructs the structural model can be analyzed. In the following, the empirical results testing the hypotheses are presented and discussed. Firstly, the structural model is evaluated by analyzing the percentage of variance explained (R^2) and the size of the structural path coefficients. In the second chapter, the prediction precision of the research model is examined analyzing the Stone-Geisser-Criterion values and the effect sizes. Finally, the total effects are analyzed their possible implications are discussed.

4.3.1 Analysis of path coefficients and variance explained

Firstly, structural models are usually evaluated by analyzing the percentage of variance explained (R^2). Secondly, the size of the structural path coefficients is discussed. Subsequently, the stability of the estimates is examined by using t-statistics obtained from jackknifing or bootstrapping resampling procedures (Venaik et al., 1999). In this analysis, the respective t-statistics are based on a bootstrapping resampling procedure (Hansmann *et al.*, 2004).

PLS is not able to provide overall goodness-of-fit measures because it has other primary research objectives than model optimization. It might even be the case that models with reasonable single goodness-of-fit statistics have to be considered poor based on other measures. The evaluation of the overall model is important for the relevance of the research model. As a rule of thumb, standardized path coefficients might be above 0.1 and ideally above 0.2 in order to be considered meaningful (Chin, 1998a; Falk *et al.*, 1992; Lohmöller, 1989). Furthermore, the percentage variance explained (R^2) needs to be interpreted based on the specific context of the endogenous and exogenous variables. According to *Backhaus, Erichson, Plinke* and *Weiber* (2003) there exist no generally applicable thresholds for acceptable R^2 values (Krafft *et al.*, 2005). The respective context needs to be integrated into the assessment of relevant R^2 values. Unfortunately, there exist no directly comparable PSM studies but an analysis of similar studies (Bartsch, 2005; Brock Smith, 1998; Dibbern *et al.*, 2005; Fritz *et al.*, 2005; Ortiz de Guinea *et al.*, 2005; Sarkar *et al.*, 2001a; Sarkar *et al.*, 2001b; Slotegraaf *et al.*, 2004; Tallon *et al.*, 2002) revealed that the achieved R^2 values are not uncommon, for example, for the percentage variance explained of company performance and PSM performance constructs. Furthermore, only two constructs are slightly below the level of 0.10 (Falk *et al.*, 1992) as required explained variance level for soft modeling approaches (Karimi *et al.*, 2004).

The supplier selection criteria constructs have R^2 values from 0.08 to 0.39 (Table 6). These results indicate that, in practice, competitive strategies and supply chain strategies still have only minor influence on the relevance of supplier selection criteria. Therefore, there probably exist other factors that influence the relevance of supplier selection criteria such as functional strategy priorities or PSM's own interest. The R^2 values for the PSM performance constructs are between 0.17 and 0.30. These figures show that the relevance of selection criteria for strategic suppliers contribute to the percentage variance explained of PSM performance. Especially, cost reductions for materials and services as well as PSM's contribution to sales increase are well explainable through the input of suppliers. Finally, the R^2 values of the company performance constructs are 0.23 for market performance and 0.17 for financial performance. Taking into account that besides PSM many other functions such as manufacturing, marketing or R&D influence company performance measures the percentage variance explained fits to the business context. Overall, the variance explained fits into the business context of the research approach.

In the following, the hypotheses stated in chapter 3.3.3 and the respective path coefficients are discussed. Accordingly, the next sections are organized following the three main hypotheses (A-C) as indicated in figure 41:

Hypothesis A: *Companies' strategic priorities in the form of competitive and supply chain strategies influence the relevance of selection criteria for strategic suppliers.*

Hypothesis B: *The relevance of selection criteria for strategic suppliers influences four strategic PSM performance levers.*

Hypothesis C: *PSM's strategic performance levers have an impact on companies' market and financial performance measures.*

The following structural path coefficients can be interpreted as standardized beta coefficients as applied in regression analyses. Figure 41 provides an overview of all hypotheses in the structural model.

Figure 41: Detailed hypotheses model

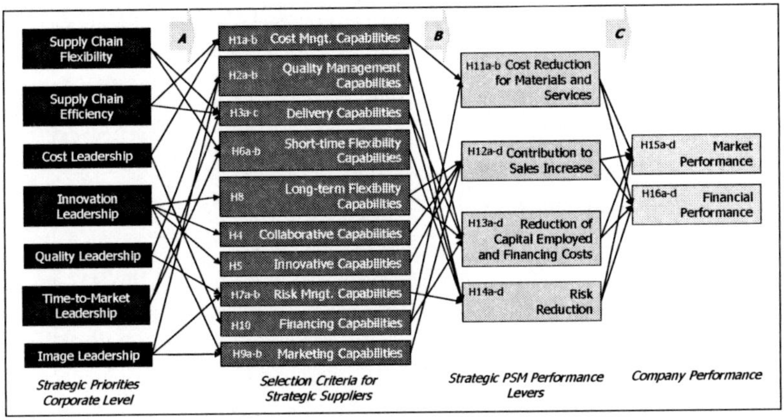

***Hypothesis A:** Companies' strategic priorities in the form of competitive and supply chain strategies influence the relevance of selection criteria for strategic suppliers.*

Cost leadership has a strong influence (0.37) on the cost management criterion while the hypothesized influence of supply chain efficiency could not be supported. Companies following a quality leadership strategy rate the quality management criterion as important (0.31). Companies following an image leadership strategy do the same but less intensive (0.19) and only significant at $p < 0.10$. The delivery capability criterion is mostly influenced by a supply chain flexibility strategy (0.40). Nevertheless, time-to-market leadership (0.29) and supply chain efficiency (0.20) also have significant influence.

The relevance of strategic suppliers' innovative capabilities is positively influenced (0.36) by a company's innovation leadership strategy. The same result is true for a strategic supplier's collaborative capabilities (0.37).

Table 6: Path coefficients and R^2 values

Hypotheses	Path Coefficients	R^2	Hypotheses	Path Coefficients	R^2
H1a Cost Leadership ⇨ Cost Management	0.37***	0,15	H12a Innovative Capabilities ⇨ Sales Increase	0.23**	0,30
H1b Supply Chain Efficiency ⇨ Cost Management	0.10^{n}		H12b Long-term Flexibility ⇨ Sales Increase	0.22**	
H2a Quality Leadership ⇨ Quality Management	0.31***	0,15	H12c Marketing Support ⇨ Sales Increase	0.14^{n}	
H2b Image Leadership ⇨ Quality Management	0.19*		H12d Collaborative Capabilities ⇨ Sales Increase	0.26**	
H3a Time-to-Market Leadership ⇨ Delivery Capabilities	0.29***	0,39	H13a Long-term Flexibility ⇨ Reduction Capital Employed	0.06^{n}	0,23
H3b Supply Chain Efficiency ⇨ Delivery Capabilities	0.20**		H13b Delivery Capabilities ⇨ Reduction Capital Employed	0.25***	
H3c Supply Chain Flexibility ⇨ Delivery Capabilities	0.40***		H13c Financing Support ⇨ Reduction Capital Employed	0.34***	
H4 Innovation Leadership ⇨ Collaborative Capabilities	0.37**	0,12	H14a Short-term Flexibility ⇨ Risk Reduction	-0.09^{n}	0,17
H5 Innovation Leadership ⇨ Innovative Capabilities	0.36***	0,11	H14b Risk Management ⇨ Risk Reduction	0.28**	
H6a Time-to-Market Leadership ⇨ Short-term Flexibility	0.31**	0,24	H14c Quality Management ⇨ Risk Reduction	0.19**	
H6b Supply Chain Flexibility ⇨ Short-term Flexibility	0.26**		H14d Delivery Capabilities ⇨ Risk Reduction	0.12^{n}	
H7a Quality Leadership ⇨ Risk Management	0.17^{n}	0,11	H15a Cost Reduction ⇨ Market Performance	0.26***	0,23
H7b Image Leadership ⇨ Risk Management	0.27**		H15b Sales Increase ⇨ Market Performance	0.24***	
H8 Innovation Leadership ⇨ Long-term Flexibility	0.29**	0,08	H15c Reduction of Capital Employed ⇨ Market Performance	0.03^{n}	
H9a Innovation Leadership ⇨ Marketing Support	0.27*	0,08	H15d Risk Reduction ⇨ Market Performance	0.18**	
H9b Image Leadership ⇨ Marketing Support	0.04^{n}		H16a Cost Reduction ⇨ Financial Performance	0.16*	0,17
H10 Cost Leadership ⇨ Financing Support	0.39***	0,13	H16b Sales Increase ⇨ Financial Performance	0.18**	
H11a Cost Management ⇨ Cost Reduction	0.36***	0,27	H16c Reduction of Capital Employed ⇨ Financial Performance	0.03^{n}	
H11b Financing Support ⇨ Cost Reduction	0.30***		H16d Risk Reduction ⇨ Financial Performance	0.24**	

* significant at 0.10 ** significant at 0.05 *** significant at 0.01 (one-tailed) ^{n}: not significant

The selection criterion short-term flexibility is mostly influenced by a company's time-to-market leadership strategy (0.31) and supply chain flexibility strategy (0.26). The importance of the risk management criterion is partly determined by the relevance of the image leadership strategy (0.27) for a company while the hypothesized impact of the quality leadership strategy could not be supported. According to the structural model evaluation, long-term flexibility is influenced by a company's innovation leadership strategy (0.29).

The selection criterion marketing support is influenced by a company's innovation leadership strategy (0.27) at $p < 0.10$. The hypothesized relationship between image leadership strategy and marketing support could not be supported. Finally, cost leadership strategy is shown to have a strong impact (0.39) on the financing support criterion.

In summary, the influence of companies' strategic priorities in the form of competitive and supply chain strategies on the relevance of selection criteria for strategic suppliers can be confirmed.

Hypothesis B: *The relevance of selection criteria for strategic suppliers influences four strategic PSM performance levers.*

Companies achieving cost reductions for materials and services through PSM focus on cost management (0.36) and financing support (0.30) as selection criteria for strategic suppliers as both are shown to significantly influence PSM's cost reduction performance.

Furthermore, the strategic supplier selection criteria innovative capabilities (0.23), long-term flexibility (0.22) and collaborative capabilities (0.26) are shown to impact PSM's contribution to sales increase. The link between marketing support and PSM's contribution to sales increase is much lower and insignificant.

Next, PSM's reduction of capital employed and financing costs is positively influenced by strategic suppliers' financing support (0.34) and delivery capabilities (0.25) while the hypothesized relationship between this third PSM performance measure and the long-term flexibility criterion could not be supported.

Finally, PSM's risk reduction performance is mostly (0.28) determined by the risk management selection criterion followed by the quality criterion (0.19). Short-term flexibility and delivery reliability have no significant impact on PSM's risk reduction performance.

In summary, the influence of selection criteria for strategic suppliers on four strategic PSM performance levers can be confirmed.

Hypothesis C: *PSM's strategic performance levers have an impact on companies' market and financial performance measures.*

A company's market performance is mostly influenced by PSM's cost reduction performance (0.26) while PSM's contribution to sales increase (0.24), and risk reduction (0.18) have slightly lower impact. Its ability to reduce capital employed and financing costs has no significant impact on market performance. A company's financial performance is mostly determined by PSM's risk reduction performance (0.24) followed by its contribution to sales increase (0.18). PSM's cost reductions activities for materials and services have slightly less (0.16) impact at $p < 0.10$. Similarly to market performance, PSM's activities to reduce a company's capital employed and financing costs could not be shown to have a significant impact on financial performance.

In summary, the impact of PSM's strategic performance levers on companies' market and financial performance measures can be confirmed.

4.3.2 Analysis of predictive relevance and effect sizes

An important aspect of PLS is its prediction accuracy. This accuracy is measured with the non-parametric Stone-Geisser-Test criterion (Geisser, 1975; Stone, 1975) applying a blindfolding procedure (Krafft *et al.*, 2005). It shows how well the empirically raised data can be reconstructed with the support of the structural model and the PLS parameters (Fornell *et al.*, 1994).

The formal calculation of the Stone-Geisser-Test criterion is as follows (Krafft *et al.*, 2005):

$$\text{Stone-Geisser-Test Criterion } Q^2_j = 1 - \frac{\Sigma_k E_{jk}}{\Sigma_k O_{jk}} > 0$$

E_{jk} = square sum of prediction errors

O_{jk} = square sum of difference between predicted value and value based on remaining data

If the value of Q^2 is above zero prediction accuracy of the structural model can be confirmed (Chin, 1998b; Fornell *et al.*, 1994). In the following, the Q^2 values for the company performance and strategic PSM performance constructs are provided (Table 7).

Table 7: Stone-Geisser-Test criterion values

Strategic PSM Performance Levers / Company Performance	
Construct	**Stone-Geisser Criterion (Q^2)**
Cost Reduction for Materials and Services	0.16 > 0
Contribution to Sales Increase	0.17 > 0
Reduction of Capital Employed and Financing Costs	0.15 > 0
Risk Reduction	0.23 > 0
Financial Performance	0.14 > 0
Market Performance	0.16 > 0

All Q^2 values are above the required zero-level and show values also reported in other studies (e.g. Ortiz de Guinea *et al.*, 2005). Although the variance explained for PSM performance construct risk reduction is not the highest it has the most prediction accuracy in the structural model. Nevertheless, the Q^2 values support the assumption that all performance constructs are almost equally well predicted through the structural model. In summary, the overall prediction accuracy of the structural model can therefore be confirmed.

Another goodness-of-fit criterion is the effect size f^2 developed by *Cohen* (1988). It measures the impact of a construct on the respective dependent variables. Its formal calculation is as follows (Krafft *et al.*, 2005):

$$\text{Effect Size } f^2 = \frac{R^2_{incl} - R^2_{excl}}{1 - R^2_{incl}} > 0$$

R^2_{incl} = R^2 values including the respective exogenous variable
R^2_{excl} = R^2 values excluding the respective exogenous variable

Effect size f^2 values of 0.02, 0.15 and 0.35 are defined as small, moderate and large effects of a construct on the respective endogenous variables respectively (Chin, 1998b; Cohen, 1988). It is important to understand that small and moderate effects are very common and do not necessarily imply unimportant effects (Chin *et al.*, 2003). The following effect size f^2 values do not completely mirror the results from the previous path coefficient analysis because, for example, an insignificant path coefficient does not imply any influence on the respective endogenous variable in terms of variance explained. The achieved values are similar to those reported in other studies (e.g. Gerpott *et al.*, 2005; Karimi *et al.*, 2004; Ortiz de Guinea *et al.*, 2005; Sanchez-Franco *et al.*, 2005; Sarkar *et al.*, 2001a). The effect size analysis of this doctoral thesis provides values for the relevant constructs from weak to above moderate and provides therefore similar results as other studies. In the following, the relevant effect size f^2 results of the structural model are described.

As expected, the variance explained of the cost management selection criterion is moderately influenced by a cost leadership strategy resulting f^2 of 0.15. A quality leadership strategy has a weak-to-medium effect size f^2 of 0.06 on the quality management criterion. The variance explained of the delivery capabilities criterion is mostly affected (0.16) by a supply chain flexibility priority on corporate level revealing an above moderate effect size f^2. The effect size f^2 of an innovation leadership priority on the collaborative capabilities criterion can be classified as moderate (0.13). The same result occurs for the influence of an innovation leadership priority on the innovative capabilities criterion (0.13). This implies that an innovation leadership impacts the variance explained of these two selection criteria to the same extent. The variance explained of the short-term flexibility criterion is slightly more influenced by the supply chain flexibility (0.07) priority than time-to-market leadership with an effect size f^2 of 0.06. The risk management criterion is mostly influenced by the image leadership strategy with an effect size f^2 of 0.08. The selection criterion marketing support is only influenced by companies with a strong innovation leadership. The effect size f^2 is 0.08. Finally, the effect size f^2 of a cost leadership strategy on the variance explained of the financing support selection criteria is 0.15.

The variance explained of the strategic PSM performance construct cost reduction for materials and services is moderately (0.17) influenced by the relevance of the supplier selection criterion cost management. The effect size f^2 of the financing support criterion is 0.10. The variance explained of the strategic PSM performance construct contribution to sales increase is mostly influenced by companies' focus on the collaborative capabilities selection criterion (0.10). The selection criterion innovative capabilities has an effect size f^2 of 0.06 and is therefore significantly less influential than the collaborative capabilities criterion. This might shed some light on the importance of those two criteria concerning sales increase. While the path coefficients are 0.26 (collaborative capabilities) and 0.23 (innovative capabilities) respectively the effect size f^2 values differ more. This implies that strategic suppliers' collaborative capabilities might indeed influence PSM's contribution to sales increase more than their innovative capabilities. The effect size f^2 of the financing support selection criteria shows a moderate influence (0.12) on the variance explained of PSM's performance in reducing capital employed and financing costs. Finally, the variance explained of PSM's risk reduction performance is mostly influenced by the selection criterion risk management (0.06).

The variance explained of companies' market performance is mostly influenced by both strategic PSM performance levers cost reductions for materials and services and contribution to sales increase with equal weak-to-moderate effect size f^2 values of 0.06. The most influential strategic PSM performance lever concerning variance explained of financial performance is risk reduction (0.06).

In summary, the Stone-Geisser-Criterion and effect size f^2 values indicate prediction accuracy and confirm the influence of the different constructs on the variance explained of the respective dependent variables. In combination with the satisfying R^2 values and the significant path coefficients, conclusions for PSM research and practice can be drawn.

4.3.3 Analysis of study results

The foregoing chapters confirmed the goodness-of-fit of the measurement models as well as the structural model. The goodness-of-fit indicators showed satisfying values in all areas. Therefore, it is possible to analyze the implications of the study results for research and practice.

As presented in chapter 1.2, this doctoral thesis focuses on the examination of the different relationships between companies' strategic priorities, the relevance of selection criteria for strategic suppliers, strategic PSM performance levers and company performance measures. The hypothesized relationships between the different constructs have been based on an extensive literature review. The proposed relationship between companies' strategic priorities and supplier selection criteria has already been presented in several articles. Similarly, the proposed relationship between the relevance of supplier selection criteria and performance measures has been identified before. Therefore, the most interesting aspects of this empirical study have been the differences between the relationships on a detailed level. The research question was not primarily whether there is any relationship between companies' strategic priorities and the relevance of selection criteria for strategic suppliers but rather which kind of influence has, for example, supply chain flexibility on the selection criterion delivery capabilities compared to supply chain efficiency. Following this approach, the study provides researchers with valuable insights about the manifold relationships between the applied constructs.

Analyzing the first main hypothesis (A), *Watts*, *Kim* and *Hahn's* model (1995) concerning the influence of companies' strategic priorities on PSM activities such as supplier selection can partly be confirmed. The low to moderate variance explained basically proves the hypothesized relationship but gives also raise to the question which other factors beside companies' strategic priorities influence the relevance of selection criteria for strategic suppliers. The model of this doctoral thesis if further supported by the path coefficient results between companies' strategic priorities and the proposed strategic PSM performance levers. For example, companies following a cost leadership strategy can be shown to perform better in PSM's cost and capital employed reduction efforts than other companies. The same applies to companies following an innovation leadership strategy and PSM's sales increase activities. Also PSM's risk reduction performance can be shown to be preferred by specific strategic priorities on corporate level such as quality leadership and image leadership.

Analyzing the second main hypothesis (B), the influence of selection criteria for strategic suppliers on strategic PSM performance levers can be confirmed. This result is similar to the study of *Vonderembse* and *Tracey* (1999) who have related supplier selection criteria to manufacturing performance. To the author's best knowledge, the applied strategic PSM performance levers so far have not been tested in a large-scale study. The reflective strategic PSM performance constructs provide a new, empirically-approved way to evaluate PSM's strategic performance. Further research might apply the proposed indicators in other large-scale PSM studies and contributes to the further refinement of these strategic PSM performance constructs. Moreover, the study has shown that the range of relevant supplier selection criteria is broader than most studies have integrated so far. For example, the financing support and long-term flexibility constructs have not been used in supplier selection criteria studies so far although they have been shown to contribute to PSM's strategic performance levers.

Analyzing the third main hypothesis (C), the influence of PSM performance on market performance and financial performance can be confirmed. As a result of this study, three out of four strategic PSM performance levers can be shown to significantly contribute to companies' market and financial performance. Therefore, this study is a further proof of PSM's strategic relevance supporting other authors' theoretical papers, e.g. from *Ramsay* (2001b) or *Mol* (2003).

The study results also provide interesting insight into PSM practice. The applied PLS approach even enables PSM managers to understand which indicators of each supplier selection construct are really driving the strategic PSM performance levers as measured in this study. Furthermore, the research methodology provides valuable insight into which strategic PSM performance levers have to be strengthened to either improve market performance or financial performance. In the following sections, the implications of the three main hypotheses are discussed.

Analyzing the first main hypothesis (A), it can be concluded that PSM managers' supplier selection criteria weighting does not necessarily represent the companies' main strategic priorities. The study results imply that other factors such as functional strategies and other interests also have a significant impact. This might reveal some room for improvement. Only if PSM is seen as a function which is able to independently provide competitive advantages for the company it might be recognized as peer from marketing, production or finance. PSM has to primarily implement a company's strategic priorities and secondary support the other functions as its internal customers.

Nevertheless, the study also shows which strategic priorities have a significant influence on the different supplier selection criteria. All proposed competitive and supply chain priorities are shown to have significant path coefficients. PSM managers should take these results into account when trying to implement a company's strategic priorities with the support of their supply base.

Analyzing the second main hypotheses (B), PSM managers can identify the most relevant single supplier selection criteria supporting the four strategic PSM performance measures. Firstly, PSM managers focusing on further cost reductions for materials and services preferably select strategic suppliers with strong capabilities in process cost management, cash-management optimization and corporate governance compliance. These results do not imply that competitive prices, for instance, are not important. The study results only show that the proposed capabilities contribute more to the generation of competitive advantages as seen in market of financial performance. Non price-competitive suppliers are never an option but it seems that successful companies focus more on process costs minimization than only on competitive prices.

Secondly, PSM managers who want to improve PSM's contribution to sales increase prioritize the following capabilities in their strategic supplier selection. They select suppliers willing to jointly develop new products or services. This requires collaborative capabilities. Based on the study results, information and knowledge exchange as well as fast conflict procedures are the most important collaborative capabilities. Suppliers' adaptation to customers' shop-floor working culture even has a negative influence on PSM's performance. This might be due to the fact that companies with the same cultural working behavior can learn less from each other than different working cultures. Finally, the study shows that suppliers' long-term flexibility in the form of geographical flexibility and product/service portfolio flexibility also contributes to sales increase.

Thirdly, PSM managers which want to reduce capital employed and financing costs prefer strategic suppliers with strong capabilities in delivery time compliance, slot size optimization, cash-management optimization and corporate governance compliance. Delivery time compliance and slot size optimization allow companies the reduction of capital employed through the transfer of inventory costs to their strategic suppliers. Additionally, cash-management optimization and corporate governance compliance support through strategic suppliers enables companies to reduce their financing costs.

Fourthly, PSM managers focusing on risk reduction preferably select strategic suppliers with strong capabilities product/service quality management and operational risk management. Financial risk management through suppliers has no significant impact on PSM's risk reduction performance. Furthermore, it is interesting to see that the impact minimization of realized risks is more important than risk prevention. This might be an indicator that risk is an inherent factor in today's business and companies are more successful if their strategic suppliers know how to handle realized risks than not to take any risks.

These detailed results are possible through the formative construction of the selection criteria constructs for strategic suppliers as proposed by *Ulaga* (2003). PSM managers might apply the study results to optimize the weighting of their selection criteria for strategic suppliers in order to optimally support their strategic priorities.

Analyzing the third main hypotheses (C), PSM managers can draw detailed conclusions about which strategic PSM performance lever to prioritize based on the CEO's expectations. For example, if a company's main objective is market growth PSM's cost reductions for materials and services as well as its contribution to sales increase have to be prioritized. If financial performance improvement is the main objective PSM's risk reduction is of pivotal importance according to the data. PSM's cost reduction efforts are shown to have only a minor influence on the generation of competitive advantages. At first sight, this result is counter-intuitive to the expected impact of cost reductions on profits in the bottom line. The explanation for this result includes two possible approaches. Firstly, many PSM functions are reducing costs since ever and the performance differences among the PSM functions of different companies might be not substantial and therefore not leading to competitive *advantages*. Secondly, PSM's achieved cost savings might not end up in the bottom line because other functions spend the saved money on other issues or the companies have to transfer achieved cost savings to the company's customers to stay in the market.

4.4 Interim results: Detailed insights into the relationships between strategic priorities, supplier selection criteria, PSM performance levers and company performance

Based on a theoretically well-grounded research model, the methodology approach included a large-scale survey analyzed with PLS. This methodology was chosen due to the missing detailed knowledge of the specific relationship between the used constructs and the possibility to simply integrate formative latent variables.

The large-scale study was pre-tested for content validity with several academics and practitioners and resulted in 112 valid questionnaires from 550 contacted companies randomly drawn from a commercial address database. The study sample represents companies from a wide range of industries and sales figures in Germany and Switzerland. The data of the responding companies support the proposition that suppliers are often responsible for more than 50% of companies' value added.

The analysis of the reflective and formative measurement models resulted in a complete confirmation of the goodness-of-fit for the applied constructs competitive strategies, supply chain strategies, supplier selection criteria, PSM performance and company performance. While the latent variables to measure the competitive strategies were already well tested in former large-scale surveys, the construction of the supply chain strategy priorities was less supported by former research. Nevertheless, both supply chain strategy priority variables fulfilled the goodness-of-fit evaluations and might serve as a reference for further studies as well. In the same way, the construction of the company performance measures was backed-up by many former studies applying financial and market performance variables while direct PSM performance measures were more difficult to find. Even so, the examination of the goodness-of-fit for the applied PSM performance constructs fulfilled all evaluation criteria. Furthermore, the use of formative supplier selection criteria proved to be successful as all evaluation criteria were fulfilled.

Based on a proper measurement model, the evaluation of the structural model provides detailed insights into the different relationships within the research model.

Firstly, the hypothesized influence of companies' strategic priorities on the relevance of selection criteria for strategic suppliers could be confirmed due to significant path coefficients. In some cases, the low variance explained serves as an indicator that many other factors such as functional priorities exist which influence the relevance of the selection criteria.

Secondly, the hypothesized influence of supplier selection criteria on PSM performance measures could be supported. Although not each single hypothesis could be confirmed the influence of the relevance of selection criteria for strategic suppliers on four PSM performance measures could be shown. Due to the fact that each of the four PSM performance measures was designed to measure overall PSM performance in its respective area the variance explained is more than satisfying.

Thirdly, the hypothesized influence of the four PSM performance measures on financial and market performance could only be partly confirmed. This is why the path coefficients of PSM's cost reduction for materials and services, contribution to sales increase and risk reduction show significant values but not the path coefficients of PSM's reduction of capital employed and financing costs. Nevertheless, the variance explained is satisfying taking PSM's or any other function's influence on company performance into account.

In summary, the applied large-scale study supports well the hypothesized research model. It provides detailed empirical knowledge about the various relationships between strategic priorities, supplier selection criteria, strategic PSM performance levers and company performance measures. PSM managers might use the research model to optimize the impact of their supply base on PSM and company performance. PSM researchers might further develop the applied strategic PSM performance levers and supplier selection criteria.

5 Summary

In the following sections, the main ideas, conceptual developments and empirical results of this doctoral thesis are summarized.

(1) Purchasing and supply management (PSM) is increasingly used as a comprehensive term which includes all activities necessary for a company to generate competitive advantages through its supply base.

Over the last thirty years, clerical purchasing has slowly developed to strategic supply management. Other terms such as procurement, materials management or sourcing are also used to describe different development stages but the terms purchasing and supply management (PSM) are increasingly used to describe the first clerical stage and the final strategic stage in research and practice. Therefore, purchasing and supply management (PSM) is increasingly used as a term which integrates all clerical and strategic activities necessary for the generation of competitive advantages through a supply base.

(2) Market-based view supports the proposition that PSM can generate competitive advantages especially in the form of monopoly rents.

The generation of monopoly rents requires the build-up of entry barriers and market power. In both cases, PSM can proactively contribute either through the erection of down-stream entry barriers for suppliers or the build-up of market power through demand bundling, market price transparency and multiple sourcing approaches. The market-based view highlights the importance of buyer dominance towards suppliers.

(3) Resource-based view supports the proposition that PSM can generate competitive advantages especially in the form of Ricardian rents.

Ricardian rents are gained by owning or controlling valuable resources and assets. They are based on strategically important resources for specific industry sectors. Companies possessing or controlling these resources have lower average costs than their competitors. For PSM, this situation implicates the anticipation of industry trends and customer expectations, the identification of potentially valuable resources such as patents, customer data bases or market access possibilities in the supply base and finally the acquisition or control of the identified resources. The acquired or controlled resources have to be protected against imitation and resource mobility, e.g., through detailed contracts or idiosyncratic activities.

(4) Capability-based view supports the proposition that PSM can generate competitive advantages especially in the form of Schumpeterian rents.

Schumpeterian rents result from dynamic and clever managerial and entrepreneurial decisions about the right combination and coordination of resources. The capability-based view assumes that the rent generation starts at that point in time when a resource is under control of the company. While the resource-based view is focused on superior information and the finding of underestimated information about which resources are strategically valuable, the mechanisms of the capability-based approach concentrate on the coordination of the different acquired or controlled resources. PSM is expected to act more like an entrepreneur. Schumpeterian rents diminish as innovations become commonplace among competitors and the only way to stay ahead of competition is to be innovative. Therefore, PSM scans the supply market for potential partners and innovations or integrates suppliers into the joint development of new products or services.

(5) Relational view supports the proposition that PSM can generate competitive advantages especially in the form of relational rents.

Two or more companies have to work together for the generation of relational rents. According to relational view, a company's competitive advantage resides not simply within the boundaries of what it owns and controls, but also in idiosyncratic interfaces it develops with other companies such as suppliers. This indicates that companies which invest into specific relationships may realize an advantage over competing companies which are unable or unwilling to do so. This may happen in the form of investments in relation-specific assets, cultural understanding, substantial knowledge exchange or joint learning efforts. This development of complementary, but scarce resources can result in the joint creation of unique new products or lower transaction costs.

(6) Principal-agent theory illustrates the relevance of information asymmetries between PSM managers and their suppliers

Information asymmetries between PSM managers and suppliers create different situations defined as hidden characteristics, hidden intentions and hidden actions. Each situation provides useful insight for PSM managers concerning their behavior towards the supply base. Hidden characteristics highlight the relevance of flexible supply contracts based on suppliers' *ex post* measured performance. Hidden intentions show the relevance of multiple sources to prevent hold-up situations and the detailed formulation of supply contracts to minimize suppliers' opportunistic behavior possibilities. Finally, hidden actions illustrate the possibilities of risk-rent sharing agreements to positively influence suppliers' performance. The prevention of these information asymmetries needs to be taken into account when weighting the respective supplier selection criteria.

(7) Transaction cost theory illustrates relevant factors for the determination of the optimal form of organizational coordination within buyer-supplier relationships

The optimal form of organizational coordination within buyer-supplier relationships is an important part of the supplier selection process. It is influenced by two of the basic characteristics of transaction cost theory: specificity of investments and uncertainty of transactions. Transaction-specific investments are assets that are uniquely tailored to a particular exchange relationship and have low value outside the focal buyer-supplier relationship. In industrial relationships, buying firms may make investments in tooling, equipment, and organizational procedures which are uniquely tailored to the specific relationship with a single supplier. Uncertainty poses a transactional problem of a different nature. It is part of the decision environment within which transactions are taking place and refers in general to situations in which the relevant decision contingencies such as volume unpredictability cannot be spelled out *ex ante*. The more specific investments are required and the higher the uncertainty is the more appropriate are coordination, collaboration or formal integration compared to open market based buyer-supplier relationships. These aspects need to be taken into account when weighting the respective supplier selection criteria.

(8) The systems theory based new St. Gallen management model can be used as an integrated framework for PSM

The University of St. Gallen has recently further developed its famous management model. Although it has been originally conceptualized for the management on corporate level its systems theory character allows the application on a functional level. Especially the structuring forces and process perspectives as well as their interaction support the development of an integrated framework for PSM practitioners and researchers. The structuring forces consist of strategy, structure and culture as most relevant determinants for a company's overall organization and behavior. The process perspective is divided into management processes, business processes and support processes. Both dimensions serve as important elements for an integrated perspective on PSM. The focus on a process perspective as *"primus inter pares"* element in today's companies provides a starting point for the development of an appropriate PSM strategy framework which allows the break-down of corporate level priorities to specific PSM activities.

(9) PSM will only be able to independently generate competitive advantages if it is appropriately integrated into companies' strategic management processes

The recognition of PSM as a true strategic function depends on its ability to independently generate competitive advantages. Only if PSM is not seen anymore as a second-class function which supports primarily production, R&D, quality management, logistics or marketing it will be able to generate competitive advantages directly for the company. This requires the integration of PSM into companies' strategic management processes top-down and bottom-up. For this reason, PSM needs to develop performance measures on its functional level that create transparency about its value creation for the company and align the selection of strategic suppliers directly with companies' strategic requirements.

(10) PSM strategy consists of the three subsystems supply strategy, sourcing strategy and support strategy

The content of purchasing, procurement, sourcing or supply strategies has often been discussed and analyzed in literature. This doctoral thesis proposes the concept of PSM strategy to structure the strategy content perspective in PSM. Based on systems theory and in accordance with strategy concepts in marketing PSM strategy is proposed to consist of three sub strategy elements (subsystems). These elements are conceptualized as supply strategy, sourcing strategy and support strategy. Supply strategy includes all decision and planning processes necessary for PSM's integration into companies' strategic processes and the management of the PSM function. The element sourcing strategy integrates PSM's core processes supplier management, supply market management and internal customer management. Finally, support strategy consists of all processes supporting PSM's management and core processes such as e-procurement or law support.

(11) Suppliers' capabilities can be applied as selection criteria for strategic suppliers

Literature on supplier selection has recently focused on capabilities as appropriate measurement scale. Especially for strategic suppliers, the potential value comes more into focus which can hardly be seen in suppliers' currently offered products or services. Therefore, the selection of strategic suppliers might be based on the detailed analysis of their capabilities in a wide range of criteria.

(12) Strategic priorities on corporate and functional level influence the relevance of selection criteria for strategic suppliers

Strategic suppliers are proposed to support companies in their strategy development and implementation. This is only possible if strategic suppliers are selected in order to support companies' strategic priorities on corporate and functional level. Relevant strategic priorities on corporate level include, for example, competitive strategies, supply chain strategies and product-/market strategies. They serve as main determinants for companies' market positioning and value proposition models and show the required supplier capabilities. Relevant strategies for PSM on functional level are logistics strategy, production strategy, and risk management strategy, R&D strategy, marketing strategy, and financing strategy. All of them follow strategic priorities which influence PSM strategy, its sub element sourcing strategy and the relevance of selection criteria for strategic suppliers as part of the supplier strategy.

(13) Cross-functional supplier selection teams improve the effectiveness of decisions for strategic suppliers

The selection of strategic suppliers can either be done by a single function or by a cross-functional team. While efficiency in the selection process is achieved through the minimization of the involved people effectiveness in the selection process in improved through cross-functional teams. Only a cross-functional approach ensures that companies' manifold interests are integrated. Nevertheless, PSM should be the main driver of supplier selection and take over process responsibility to ensure that corporate level strategies are prioritized.

(14) A combination of different criteria weighting methods supports an efficient and objective supplier selection process

Many different criteria weighting approaches have been developed over the last decades. Until today, PSM practice has not taken over any of the recently developed sophisticated weighting approaches to a great extent. This is why only a combination of different weighting methods combines the often required efficient handling and objectivity. Therefore, a mixture of the classical weighted points method and the rather new analytical network process (ANP) method are proposed for the selection of strategic suppliers. In this way, the requirements from practice concerning easy handling (weighted points method) and necessary objectivity (ANP) are combined in different stages of the criteria weighting process.

(15) The selection of strategic suppliers requires a wide range of selection criteria

Strategic suppliers are proposed to support the development and implementation of companies' strategic priorities and generate competitive advantages. Consequently, the selection of strategic suppliers should be based on their weighted overall contribution to companies' strategy implementation. A broad range of selection criteria needs therefore to be applied in order to measure the overall contribution. As a consequence, classical selection criteria such as price, quality or delivery are still important but need to be complemented by new criteria such as marketing support, risk management or financing support. Furthermore, recently identified criteria such as collaborative capabilities or innovative capabilities need to be included. Finally, strategy-based supplier selection approaches also involve short-term and long-term flexibility.

(16) Companies' strategic priorities influence the relevance of selection criteria for strategic suppliers but their influence is in practice not as large as proposed by many PSM researchers

The conducted study provides detailed insights into the influence of companies' strategic priorities on the relevance of selection criteria for strategic suppliers. Criteria such as cost or quality management are influenced by their respective strategic priorities on corporate level. The criterion delivery capabilities is mostly influenced by supply chain flexibility. Furthermore, time-to-market leaders also focus on this selection criterion. The criteria collaborative and innovative capabilities are both strongly influenced by innovation leadership priorities. Short-term flexibility is in the focus of companies which follow a time-to-market leadership or have flexibility-oriented supply chains. Long-term flexibility is an important criterion for innovations leaders. Risk management is especially important for image leaders. The selection criterion marketing support is only in the focus of innovation leaders. Finally, the financing support criterion is strongly influenced by cost leadership priorities. Although the influence can be confirmed, the research results show that the influence of companies′ strategic priorities is not as high as theoretically proposed in earlier conceptual PSM research papers.

(17) The influence of selection criteria for strategic suppliers on four strategic PSM performance levers shows the relevance of supplier selection for companies

The conducted study provides detailed insights into the influence of supplier selection criteria on four strategic PSM performance levers. These four PSM performance levers are cost reduction for materials and services, contribution to sales increase, reduction of capital employed and financing costs and risk reduction. PSM's cost reduction efforts are supported through the selection of strategic suppliers with a high performance in cost management and financing support. PSM's contribution to sales increase is strongly supported through suppliers with excellent collaborative and innovative capabilities. Moreover, suppliers with long-term flexibility performance positively influence PSM's contribution to sales increase. Companies focusing on the reduction of capital employed and financing costs are advised to select strategic suppliers with excellent delivery capabilities and financing support. Finally, PSM's risk reduction efforts are supported through the selection of strategic suppliers with a high risk and quality management performance. Overall, a significant influence of supplier's capabilities on the proposed four strategic PSM performance levers can be confirmed.

(18) The influence of four strategic PSM performance levers on companies' market and financial performance confirms PSM's new strategic role

The conducted study provides detailed insights into the influence of four strategic PSM performance levers on companies' market and financial performance. Market performance is mostly influenced through PSM's cost reduction performance but also through its contribution to sales increase and risk reduction efforts. Financial performance is mostly influenced through PSM's risk reduction efforts. Sales increase also has a positive influence while PSM's cost reduction performance is less important.

(19) PSM is shown to contribute to the generation of competitive advantages if its supplier selection is based on companies' strategic priorities

This doctoral thesis has developed a capability-focused, strategy-based supplier selection approach which integrates a wide range of criteria. These criteria are proposed to be primarily based on companies' strategic priorities and secondary on their functions' strategic priorities. With this approach, PSM is able to support the implementation of companies' manifold strategies and generate competitive advantages in the form of four rents: monopoly rents, Ricardian rents, Schumpeterian rents, and relational rents. PSM's direct influence on company performance measures as indicators for the generation of competitive advantages can be shown through its four performance levers cost reductions for materials and services, contribution to sales increase, reduction of capital employed and financing costs, and risk reduction.

(20) PSM managers should select their strategic suppliers based on their capabilities supporting the implementation of their companies' strategic priorities

The conceptual and empirical results of this doctoral thesis show the importance of a strategy-based supplier selection to support companies' strategy implementation efforts. Only if PSM strategy is directly aligned with companies' strategic priorities it can independently generate competitive advantages and is perceived as strategic function. This direct link to corporate level priorities allows an analysis of the identified capability gap between market positioning and value proposition. A broad range of supplier selection criteria is required which focus on strategic suppliers' various long-term capabilities to support strategy implementation.

Appendix: Cross-loading table

	Innovation Leadership	Cost Leadership	Quality Leadership	Image Leadership	Time-to-Market Leadership	Supply Chain Flexibility	Supply Chain Efficiency	Cost Reduction of Materials and Services	Contribution to Sales Increase	Reduction of Capital Employed and Financing Costs	Risk Reduction	Market Performance	Financial Performance
Efficient Administration	0.27	**0.80**	0.23	0.20	0.37	0.45	0.13	0.28	0.10	0.36	0.10	0.18	0.12
Efficient Sales	0.25	**0.76**	0.05	0.31	0.54	0.39	0.09	0.43	0.00	0.44	0.09	0.20	0.18
Production Costs	0.32	**0.79**	0.11	0.21	0.49	0.42	0.25	0.38	0.00	0.37	0.21	0.11	0.15
Scale Effects	0.29	**0.72**	0.21	0.33	0.37	0.38	0.17	0.31	-0.04	0.27	0.22	0.18	0.11
Further Product Development	0.31	**0.73**	0.24	0.30	0.55	0.51	0.13	0.27	0.12	0.29	0.11	0.13	0.12
Efficient Capacity Management	0.17	**0.77**	0.17	0.25	0.47	0.38	0.28	0.28	0.03	0.33	0.04	-0.09	-0.04
Innovation Developement	**0.84**	0.24	0.30	0.24	0.19	0.27	0.10	0.07	0.17	-0.04	0.22	0.23	0.20
Use of Technological Knowledge	**0.81**	0.33	0.40	0.31	0.29	0.38	0.02	0.14	0.22	0.15	0.19	0.21	0.22
New Technology Integration	**0.79**	0.30	0.36	0.12	0.29	0.23	0.17	0.16	0.31	0.09	0.25	0.23	0.22
R&D Investments	**0.87**	0.32	0.37	0.31	0.20	0.26	0.18	0.14	0.37	0.08	0.18	0.36	0.28
Innovation Management	**0.88**	0.30	0.35	0.21	0.24	0.35	0.11	0.10	0.31	0.05	0.18	0.22	0.18
Strong Brands	0.20	0.27	0.29	**0.78**	0.24	0.20	0.03	0.16	0.21	0.16	0.15	0.16	0.24
Innovative Design	0.30	0.24	0.17	**0.82**	0.38	0.28	0.14	0.12	0.20	0.13	0.25	0.20	0.18
Market Research	0.09	0.25	0.10	**0.81**	0.29	0.21	0.00	0.18	0.17	0.12	0.16	0.23	0.20
Marketing Intensity	0.33	0.38	0.19	**0.77**	0.45	0.25	0.17	0.13	0.18	0.20	0.21	0.23	0.11
Flexible Processes	0.18	0.46	0.06	0.29	**0.74**	0.45	0.05	0.36	0.17	0.30	0.29	0.12	0.15
Short Delivery Times	0.20	0.56	0.18	0.31	**0.83**	0.49	0.15	0.38	0.12	0.29	0.30	0.10	0.12
Reaction on Trends	0.23	0.47	0.01	0.38	**0.87**	0.48	-0.03	0.33	0.14	0.19	0.18	0.12	0.05
Trend Identification	0.32	0.46	0.22	0.45	**0.79**	0.42	0.00	0.25	0.14	0.13	0.21	0.22	0.12
High Performance	0.33	0.13	**0.82**	0.14	0.13	0.15	0.13	-0.06	0.08	-0.04	0.24	0.05	0.26
Zero-Mistake Production	0.33	0.16	**0.79**	0.18	0.07	0.01	0.04	-0.05	0.05	-0.08	0.05	0.12	0.26
Quality Standards	0.39	0.25	**0.89**	0.24	0.16	0.24	0.11	-0.02	0.06	-0.08	0.20	0.18	0.28
Build-to-Order Production	0.38	0.27	0.02	0.24	0.37	**0.66**	-0.02	0.31	0.16	0.31	0.21	0.23	0.14
Product Life Cycle	0.25	0.41	0.13	0.25	0.40	**0.74**	-0.21	0.27	-0.03	0.27	0.21	0.15	-0.01
Information Exchange	0.16	0.38	0.12	0.11	0.44	**0.70**	-0.05	0.34	0.02	0.31	0.30	-0.03	-0.11
Supply Shortage Costs	0.26	0.51	0.27	0.15	0.39	**0.77**	-0.04	0.19	0.09	0.16	0.14	0.03	0.09
Market Trends	0.17	0.39	0.04	0.35	0.48	**0.70**	-0.14	0.33	0.01	0.31	0.12	0.11	-0.03
Inventory Costs	0.18	0.15	0.14	0.08	0.04	-0.14	**0.92**	0.15	0.24	0.07	0.14	0.17	0.11
Transportation Costs	0.07	0.28	0.06	0.15	0.12	-0.02	**0.79**	0.09	0.07	0.17	0.18	0.00	-0.01
Transaction Costs	0.00	0.17	0.02	0.02	-0.04	-0.13	**0.62**	0.05	0.13	0.05	0.15	-0.01	-0.07
Bundling	0.18	0.32	-0.05	0.19	0.30	0.38	0.06	**0.82**	0.19	0.41	0.30	0.38	0.23
Price Negotiations	0.05	0.35	-0.02	0.11	0.39	0.36	0.05	**0.77**	0.02	0.42	0.30	0.23	0.24
Specification Management	0.11	0.33	-0.05	0.12	0.33	0.28	0.17	**0.77**	0.21	0.43	0.11	0.29	0.25
Standardization	0.10	0.29	-0.02	0.13	0.22	0.13	0.17	**0.68**	0.14	0.33	0.08	0.20	0.09
Supplier Risks	0.27	0.15	0.15	0.15	0.29	0.30	0.19	0.20	0.18	0.20	**0.85**	0.19	0.21
Supply Organization Risks	0.19	0.15	0.08	0.17	0.30	0.24	0.16	0.31	0.07	0.20	**0.82**	0.20	0.25
Supply Risks	0.15	0.13	0.17	0.23	0.26	0.18	0.16	0.18	0.12	0.10	**0.82**	0.31	0.27
Currency Risks	0.15	0.12	0.24	0.23	0.12	0.14	0.08	0.15	0.17	0.27	**0.68**	0.22	0.27
Cash Management Optimization	0.06	0.39	-0.12	0.08	0.23	0.28	0.14	0.46	0.12	**0.86**	0.19	0.15	0.12
Financing Optimization	0.05	0.40	-0.06	0.12	0.28	0.27	0.09	0.48	0.02	**0.85**	0.34	0.20	0.17
Reduction of Financing Costs	0.08	0.32	-0.02	0.13	0.16	0.26	0.07	0.32	0.20	**0.83**	0.24	0.12	0.13
Inventory Optimization	0.07	0.35	-0.05	0.25	0.24	0.37	0.08	0.43	0.18	**0.74**	0.12	0.32	0.24
Reduction of Development Times	0.18	0.06	-0.04	0.18	0.22	0.11	0.13	0.22	**0.72**	0.11	0.15	0.27	0.14
Innovation Identification	0.33	0.01	0.11	0.24	0.07	0.02	0.19	0.16	**0.91**	0.15	0.12	0.37	0.31
Innovation Development	0.24	0.02	0.00	0.16	0.12	0.05	0.21	0.15	**0.85**	0.16	0.19	0.20	0.19
Product Value Improvement	0.33	0.08	0.18	0.16	0.21	0.14	0.08	0.04	**0.61**	0.07	0.07	0.05	0.08
Product/Service Portfolio	0.32	0.03	0.29	0.10	0.09	0.09	0.07	0.19	0.26	0.01	0.18	**0.70**	0.52
Market Share	0.27	0.16	0.11	0.29	0.16	0.14	0.12	0.31	0.23	0.28	0.29	**0.91**	0.65
Sales Volume	0.22	0.18	0.04	0.25	0.17	0.11	0.08	0.40	0.32	0.29	0.26	**0.92**	0.67
Return on Investments	0.30	0.21	0.30	0.29	0.12	0.03	0.06	0.23	0.25	0.20	0.28	0.69	**0.91**
Return on Equity	0.23	0.13	0.32	0.13	0.09	0.03	0.11	0.24	0.21	0.16	0.22	0.60	**0.89**
Return of Sales	0.20	0.06	0.27	0.19	0.14	0.05	-0.01	0.26	0.24	0.21	0.35	0.68	**0.93**

References

A.T. Kearney Inc. 1998. Leveraging the Strategic Nature of Procurement. A White Paper on How the Leaders Are Creating Competitive Advantage: Chicago, IL

A.T. Kearney Inc. 2004. How Many Supply Chains Do You Need? Matching Supply Chains Strategies to Products and Customers: Chicago, IL

Aberle G, Eisenkopf A. 2002. Wettbewerbspolitische Probleme innovativer Beschaffungskonzeptionen. In D Hahn, L Kaufmann (Eds.), *Handbuch industrielles Management*, 2 ed.: 165-179. Gabler: Wiesbaden

Agrawal MK, Minsok HP. 2001. Getting smart about supply chain management. *The McKinsey Quaterly* **37**(2): 22-26

Albers S, Hildebrandt L. 2006. Methodische Probleme bei der Erfolgfaktorenforschung - Messfehler, formative versus reflektive Indikatoren und die Wahl des Strukturgleichungs-Modells. *Zeitschrift für Betriebswirtschaft* **58**: 2-33

Al-Mudimigh AS, Zairi M, Ahmed AMM. 2004. Extending the concept of supply chain: The effective management of value chains. *International Journal of Production Economics* **87**: 309-320

Amit R, Livnat J. 1988. Diversification strategies, business cycles and economic performance. *Strategic Management Journal* **9**: 99-110

Amit R, Shoemaker PJH. 1993. Strategic assets and organizational rent. *Strategic Management Journal* **14**: 33-46

Anders W. 1992. *Strategische Einkaufsplanung*. Lang: Frankfurt am Main

Andersen PH, Rask M. 2003. Supply chain management: new organisational practices for changing procurement realities. *Journal of Purchasing & Supply Management* **9**: 83-95

Anderson JC, Gerbing DW. 1991. Predicting the Performance of Measures in a Confirmatory Factor Analysis with a Pretest Assessment of their Substantive Validities. *Journal of Applied Psychology* **76**(5): 732-740

Anderson JC, Hakansson H, Johanson J. 1994. Dyadic business relationships within a business network context. *Journal of Marketing* **58**(4): 1-15

Anderson JC, Narus JA. 1999. *Business Market Management: Understanding, Creating and Delivering Value*. Prentice Hall: New Jersey

Anonymous. 1988. Group Decision Making: Approaches To Problem Solving. *Small Business Report* **13**(7): 30-33

Anonymous. 1992. How Capabilities Differ from Core Competencies. The Case of Honda. *Harvard Business Review* **70**(2): 66

Ansoff HI. 1958. A Model for Diversification. *Management Science* **4**(4): 392-414

Ansoff HI. 1966. *Management – Strategie. München.* Verlag Moderne Industrie: München

Araujo L, Dubois A, Gadde L-E. 1999. Managing Interfaces with Suppliers. *Industrial Marketing Management* **28**: 497-506

Arnold U. 1997. *Beschaffungsmanagement* (2 ed.). Schäffer-Poeschel: Stuttgart

Arnold U. 2002. Global Sourcing: Strategiedimensionen und Strukturanalyse. In D Hahn, L Kaufmann (Eds.), *Handbuch industrielles Management*, 2 ed.: 203-220. Gabler: Wiesbaden

Arnold U. 2004. Supplier Lifetime Value - Ein Konzept zur Lieferantenbewertung in Industrie und Handel. In HH Bauer, F Huber (Eds.), *Strategien und Trends im Handelsmanagement*: 177-196. Vahlen: München

Arnold U, Eßig M. 2003. The Supplier Lifetime Value Model - A Strategic Approach for Purchasing Relationship Controlling, *14th Annual North American Research Symposium on Purchasing and Supply Management*. CAPS: Tempe, AZ

Arnold U, Warzog F. 2005. Supplier Relationship Management - An Explorative Study Concentrated on the Controlling Process., *16th North American Research Symposium*: Phoenix, AZ

Babu N, Olson DO. 1989. On the Definition of Barriers to Entry. *Southern Economic Journal* **56**(1): 236-239

Backhaus K. 1999. *Industriegütermarketing* (6 ed.). Vahlen: München

Backhaus K, Erichson B, Plinke W, Weiber R. 2003. *Multivariate Analysemethoden: Eine anwendungsorientierte Einführung* (10 ed.). Springer: Berlin

Badaracco JL. 1991. *The knowledge link: How firms compete through strategic alliances.* Harvard Business School Press: Boston, MA

Bagozzi RP. 1994. Structural Equation Models in Marketing Research: Basic Principles. In RP Bagozzi (Ed.), *Principles of Marketing Research*: 317-385. Blackwell: Oxford

Bagozzi RP, Youjae Y. 1988. On the Evaluation of Structural Equation Models. *Journal of the Academy of Marketing Science* **16**(1): 74-94

Baily P. 2001. *Purchasing and Supply Management* (5 ed.). Chapman & Hall: London

Baily P, Farmer D, Jessop D, D J. 1994. *Purchasing Principles and Management* (7 ed.). Pitman Publishing: London

Bain JS. 1956. *Barriers to new competition.* Harvard University Press: Cambridge

Bain JS. 1968. *Industrial Organization.* Wiley: New York, NY

Barney JB. 1991. Firm Resources and Sustained Competitive Advantage. *Journal of Management* **17**: 99-120

Barney JB. 1996. Bringing managers back in: a resource-based analysis of the role of managers in creating and sustaining competitive advantages for firms. In JB Barney, JC Spender, T Reve (Eds.), *Does Management Matter? On Competencies and Competitive Advantage*: 1-36. Lund University Press: Lund

Barney JB. 2001. Is the resource-based "view" a useful perpective for strategic management research? Yes. *Academy of Management Review* **26**(1): 41-56

Bartsch A. 2005. *Lieferantenwert. Auswirkungen der Eigenschaften von Lieferanten auf Nutzen und Aufwand bei industriellen Kunden*. DUV: Wiesbaden

Baumgarten H, Darkow I-L. 2002. Versorgungsmanagement - Integration von Beschaffung und Logistik. In D Hahn, L Kaufmann (Eds.), *Handbuch industrielles Management*, 2 ed.: 383-399. Gabler: Wiesbaden

Beamon BM. 1999. Measuring supply chain performance. *International Journal of Operations & Production Management* **19**(3): 275-292

Becker J. 1998. *Marketing-Konzeption: Grundlagen des strategischen und operativen Marketing-Management* (6 ed.). Vahlen: München

Bellmann K. 2002. Produktion und Beschaffung - Management einer innerbetrieblichen Schnittstelle. In D Hahn, L Kaufmann (Eds.), *Handbuch industrielles Management*, 2 ed.: 364-379. Gabler: Wiesbaden

Bengtsson A, Servais P. 2004. Co-branding and the impact on inter-organizational relationships, *20th Annual IMP Conference*: Copenhagen

Bentler PM, Weeks DG. 1980. Linear structural equations with latent variables. *Psychometrika* **45**: 289-308

Bertalanaffy Lv. 1995. *General Systems Theory. Foundations, Development, Applications* (12 ed.). George Braziller: New York, NY

Bhote KR. 1989. *Strategic Supply Management. A Blueprint for Revitalizing the Manufacturer-Supplier Partnership*. American Management Association: New York, NY

Bhutta KS, Huq F. 2002. Supplier selection problem: a comparison of the total cost of ownership and analytic hierarchy process approaches. *Supply Chain Management: An International Journal* **7**(3): 126-135

Biong H, Wathne K, Parvatiyar A. 1997. Why Do Some Companies Not Want to Engage in Partnering Relationships. In HG Gemünden, T Ritter, A Walter (Eds.), *Relationships and Networks in International Business Markets*: 91-107. Pergamon: New York, NY

Birou LM, Fawcett SE, Magnan GM. 1997. Integrating Product Life Cycle and Purchasing Strategies. *International Journal of Purchasing and Materials Management* **33**(1): 23-31

Bohrnstedt GW. 1970. Reliability and Validity Assessment in Attitude Measurement. In GF Summers (Ed.), *Attitude Measurement*: 80-99. Wiley: London

Bollen KA, Lennox R. 1991. Conventional wisdom on measurement: A structural equation perspective. *Psychological Bulletin* **110**(2): 305-314

Bortz J. 1999. *Statistik für Sozialwissenschaftler* (5 ed.). Springer: Berlin

Boston Consulting Group. 2004. Beyond Cost Reductions. Report: Boston, MA

Boutellier R, Corsten D. 1997. Total Supply Management. In HP Wehrli (Ed.), *Erfolgspotentiale in der Beschaffungslogistik*: 159-175. SVME: Aarau

Bowman C, Ambrosini V. 1998. Value Creation Versus Value Capture: Towards A Coherent Definition of Value in Strategy - An Exploratory Study, *Cranfield University Working Paper*: Cranfield

Bozarth C, Handfield RB, Das A. 1998a. Stages of global sourcing strategy evolution: an exploratory study. *Journal of Operations Management* **16**: 241-255

Bozarth C, McDermott C. 1998b. Configurations in manufacturing strategy: a review and directions for future research. *Journal of Operations Management* **16**: 427-439

Bozdogan K, Deyst J, Hoult D, Lucas M. 1998. Architectural innovation in product development through early supplier integration. *R&D Management* **28**(3): 163-173

Bratzler M. 1999. Strategisches Management von Technologie-Lieferanten, *University of St. Gallen*. Dissertation: St. Gallen

Brock Smith J. 1998. Buyer-Seller Relationships: Similarity, Relationship Management, and Quality. *Psychology & Marketing* **15**(1): 3-21

Browning JM, Zabriskie NB, Huellmantel AB. 1983. Strategic purchasing planning. *Journal of Purchasing and Materials Management* **19**(1): 19-24

Buck T. 1998. *Konzeption einer integrierten Beschaffungskontrolle*. DUV: Wiesbaden

Burnes B. 1997. Collaboration In Customer-Supplier Relationships: Strategy, Operations and the Function of Rhetoric. *International Journal of Purchasing and Materials Management* **33**(3): 10-17

Burt DN, Dobler DW, Starling S. 2002. *World Class Supply Management. The Key to Supply Chain Management* (7 ed.). McGraw-Hill/Irwin: Boston, MA

Butz HE, Goodstein LD. 1996. Measuring Customer Value: Gaining the Strategic Advantage. *Organizational Dynamics* **24**: 63–77

Cannon JP, Homburg C. 2001. Buyer-supplier relationships and customer firm costs. *Journal of Marketing* **65**(1): 29-43

Carr AS, Pearson JN. 1999a. Strategically Managed Buyer-Supplier Relationships and Performance Outcomes. *Journal of Operations Management* **17**(5): 497-519

Carr AS, Pearson JN. 2002. The impact of purchasing and supplier involvement on strategic purchasing and its impact on firm's performance. *International Journal of Operations and Production Management* **22**(9): 1032-1053

Carr AS, Smeltzer LR. 1997. An empirically based operational definition of strategic purchasing. *European Journal of Purchasing & Supply Management* **3**: 199-207

Carr AS, Smeltzer LR. 1999b. The relationship of strategic purchasing to supply chain management. *European Journal of Purchasing & Supply Management* **5**: 43-51

Carter J, Narasimhan R. 1996. Is Purchasing Really Strategic? *International Journal of Purchasing and Materials Management* **32**(1): 20-28

Carter J, Smeltzer LR. 2003. Critical Strategy Areas for Purchasing and Supply Management. In World Markets Research Centre (Ed.), *Global Purchasing & Supply Chain Strategies*: 36-41. World Markets Research Centre: London

Carter J, Vickery SK, D'Itri MP. 1993. Currency Risk Management Strategies for Contracting with Japanese Suppliers. *International Journal of Purchasing and Materials Management* **29**(3): 19-25

Carter PL, Monczka RM, Mosconi T. 2005. *Strategic Performance Measurement for Purchasing and Supply*. CAPS: Tempe, AZ

Carter PL, Monczka RM, Slaight TH, Swan AJ. 1998. *The Future of Purchasing and Supply: A Five- and Ten-Year Forecast*. CAPS: Tempe, AZ

Cassel CM, Hackl P, Westlund AH. 1999. Robustness of partial least squares method for estimating latent variable quality structures. *Journal of Applied Statistics* **26**: 435-446

Cavinato JL. 1999. Fitting purchasing to the five stages of strategic management. *European Journal of Purchasing & Supply Management* **5**: 75-83

Cavinato JL. 2000. Uncover the Organization's Financial Layers, *Purchasing Today*: 18-20

Cavinato JL. 2003. Sarbanes-Oxley and the Supply Chain, *Inside Supply Management*, Vol. 14: 26

Cavinato JL, Kauffman RG. 2000. *The Purchasing Handbook. A Guide for the Purchasing and Supply Professional*. McGraw-Hill: New York, NY

Chen C-C, Yeh T-M, Yang C-C. 2004. Customer-focused rating system of supplier quality performance. *Journal of Manufacturing Technology Management* **15**(7): 599-606

Chin WW. 1998a. Issues and Opinion on Structural Equation Modeling. *MIS Quaterly* **22**(3): VII-XVI

Chin WW. 1998b. The partial least squares approach to structural equation modelling. In GA Marcoulides (Ed.), *Modern methods for business research*: 295-336. Lawrence Erlbaum Associates: Mahwah

Chin WW, Marcolin BM, Newsted PR. 2003. A Partial Least Squares Latent Variable Modeling Approach for Measuring Interaction Effects: Results from a Monte-Carlo Simulation Study and an Electronic-Mail Emotion / Adoption Study. *Information Systems Research* **14**(2): 189-217

Chin WW, Newsted PR. 1999. Structural equation modeling analysis with small samples using partial least squares. In RH Hoyle (Ed.), *Statistical strategies for small sample research*: 307-342. SAGE: Thousand Oaks, CA

Chou S. 2005. The Sarbox Effect, *European Leaders in Procurement*, Vol. 1: 28-31

Christopher M. 1992. *Logistics and Supply Chain Management*. Pitman Publishing: London

Christopher M. 1994. *Logistics and Supply Chain Management. Strategies for Reducing Costs and Improving Services*. Prentice Hall: London

Christopher M. 1999. Supply chain strategy: its impact on shareholder value. *International Journal of Logistics Management* **10**(1): 1-10

Christopher M, Jüttner U. 2000. Developing strategic partnerships in the supply chain: a practitioner perspective. *European Journal of Purchasing & Supply Management* **6**: 117-127

Christopher M, Lee H. 2004. Mitigating supply chain risk through improved confidence. *International Journal of Physical Distribution & Logistics Management* **34**(5): 388-396

Chung S, Harbir S, Lee K. 2000. Complementarity, status similarity and social capital as drivers of alliance formation. *Strategic Management Journal* **21**: 1-22

Chung WWC, Yam AYK, Chan MFS. 2004. Networked enterprise: A new business model for global sourcing. *International Journal of Production Economics* **87**: 267-280

Claus H, Betts B. 2003. *Adapt or Die. Transforming Your Supply Chain into an Adaptive Business Network*. Wiley: Hoboken, NJ

Clouse M, Busch J. 2003. Global Supply Management: Strategies for Identifying and Managing Supply Risk: San Francisco, CA

Coase R. 1937. The Nature of the Firm. *Economica* **4**: 386-405

Cohen J. 1988. *Statistical Power Analysis for the Behavioral Sciences* (2 ed.). Lawrence Erlbaum Associates: Hillsdale, NJ

Collis DJ, Montgomery CA. 1995. Competing on Resources: Strategy in the 1990s. *Harvard Business Review* **73**(4): 118-128

Colsman PG. 2000. *Global Sourcing als eine Beschaffungsstrategie für globale Unternehmen*. Fördergesellschaft Produktmarketing e.V.: Köln

Contractor FJ, Kim C, Beldona S. 2002. Interfirm learning in alliances and technology networks: An empirical study in the global pharmaceutical and chemical industries. In FJ Contractor, P Lorange (Eds.), *Cooperative strategies and alliances*: 493-516. Pergamon: Amsterdam

Cook WD, Johnston DA. 1992. Evaluating Suppliers of Complex Systems: A Multiple Criteria Approach. *Journal of Operational Research Society* **43**: 1055-1061

Cooper M, Lambert DM, Pagh JD. 1997. Supply Chain Management: More than a New Name for Logistics. *The International Journal of Logistics Management* **8**: 1-14

Copacino WC, Anderson D. 2004. A Global Study of Supply Chain Leadership and Its Impact on Business Performance: Chicago, IL

Correia G, Neto JA. 2005. Supplier Selection as a Strategic Competency of Purchasing: A Transaction Cost Economics Approach, *16th North American Research Symposium*: 139-158: Phoenix, AZ

Cousins PD. 2002. A conceptual model for managing long-term inter-organisational relationships. *European Journal of Purchasing & Supply Management* **8**: 71-82

Cousins PD. 2005. The alignment of appropriate firm and supply strategies for competitive advantage. *International Journal of Operations & Production Management* **25**(5): 403-428

Cousins PD, Spekman R. 2003. Strategic supply and the management of inter- and intra-organisational relationships. *Journal of Purchasing & Supply Management* **9**: 19-29

Cox A, Hughes EJ, Ralf M. 1995. Influencing the strategic agenda: The challenge for purchasing leadership, *Purchasing & Supply Management*: 36-41

Cox A, Lamming R. 1997. Managing supply in the firm of the future. *European Journal of Purchasing & Supply Management* **3**: 53-62

Cravens DW, Greenley G, Piercy N, Slater S. 1997. Integrating Contemporary Strategic Management Perspectives. *Long Range Planning* **30**(4): 493-506

Croom S, Batchelor J. 1997. The development of strategic capabilities. An interaction view. *Integrated Manufacturing Systems* **8**(5): 29-312

Cuclos LK, Vokurka RJ, Lummus RR. 2003. A conceptual model of supply chain flexibility. *Industrial Management & Data Systems* **103**(6): 446-456

Culley SJ, Boston OP, McMahon CA. 1999. Suppliers in New Product Development: Their Information and Integration. *Journal of Engineering Design* **10**(1): 59-75

Cummings S, Angwin D. 2004. The future shape of strategy: Lemmings or chimeras? *Academy of Management Executive* **18**(2): 21-36

Curtis RF, Jackson EF. 1962. Multiple Indicators in Survey Research. *American Journal of Sociology* **68**(2): 195-204

Das A, Narasimhan R. 2000. Purchasing Competence and Its Relationship with Manufacturing Performance. *Journal of Supply Chain Management* **36**(2): 17-28

D'Avanzo R, von Lewinski H, van Wassenhove LN. 2003. The Link Between Supply Chain and Financial Performance. *Supply Chain Management Review* **7**(6): 40-47

de Boer L, Harink J, Heijboer G. 2001a. A model for assessing the impact of electronic procurement forms, *11th IPSERA Conference*: Enschede

de Boer L, Labro E, Morlacchi P. 2001b. A review of methods supporting supplier selection. *European Journal of Purchasing & Supply Management* **7**: 75-89

de Boer L, van der Wegen L, Telgen J. 1998. Outranking methods in support of supplier selection. *European Journal of Purchasing & Supply Management* **4**: 109-118

de Quervain MA, Wagner SM. 2003. Von der Strategiefindung zur Strategieumsetzung. In R Boutellier, SM Wagner, HP Wehrli (Eds.), *Handbuch Beschaffung. Strategien, Methoden, Umsetzung*: 99-131. Hanser: München

De Toni A, Nassimbeni G. 2000. Just-in-time purchasing: an empirical study of operational practices, supplier development and performance. *Omega* **28**: 631-651

Dess GG, Davis PS. 1984. Porter's (1980) Generic Strategies as Determinants of Strategic Group Membership and Organizational Performance. *Academy of Management Journal* **27**(3): 467-488

DeVellis RF. 1991. *Scale Development: Theory and Applications*. Sage Publications: Newbury Park, CA

Diamantopoulos A, Winklhofer HM. 2001. Index Construction with Formative Indicators: An Alternative to Scale Development. *Journal of Marketing Research* **38**: 269-277

Dibbern J, Chin WW. 2005. Multi-Group Comparison: Testing a PLS Model on the Sourcing of Applications Software Services across Germany and the U.S.A. Using a Permutation Based Algorithm. In F Bliemel, A Eggert, G Fassott, J Henseler (Eds.), *Handbuch PLS-Pfadmodellierung. Methode, Anwendung, Praxisbeispiele*: 135-160. Schäffer-Poeschel: Stuttgart

Dickson GW. 1966. An Analysis of Supplier Selection Systems and Decisions. *Journal of Purchasing* **2**: 5-17

Dierickx I, Cool K. 1989. Asset stock accumulation and sustainability of competitve advantage. *Management Science* **35**(12): 1504-1511

Disney SM, Potter AT, Gardner BM. 2003. The impact of Vendor Managed Inventory on transport operations. *Transportation Research Part E* **39**(5): 363-380

Disselkamp M, Schüller R. 2004. *Lieferantenrating. Instrumente, Kriterien, Checklisten.* Gabler: Wiesbaden

Dobler DW, Burt DN. 1996. *Purchasing and Supply Management. Text and Cases* (6 ed.). McGraw-Hill: New York, NY

Dobler DW, Lee L. 1990. *Purchasing and Materials Management. Text and Cases* (5 ed.). McGraw-Hill: New York, NY

Doyle MF. 1990. Fundamentals of strategic supply management. *Purchasing World* **32**(12): 40-41

Drucker P. 1982. *The Changing World of the Executive*. Times Book: New York, NY

Dubois A, Hulthén K, Pedersen A-C. 2004. Supply chains and interdependence: a theoretical analysis. *Journal of Purchasing & Supply Management* **10**: 3-9

Duffy RJ. 1999. The Future of Purchasing and Supply: Supply Chain Partner Selection and Contribution, *Purchasing Today*: 41

Dulmin R, Mininno V. 2003. Supplier selection using a multi-criteria decision aid method. *Journal of Purchasing & Supply Management* **9**: 177-187

Duschek S. 2004. Inter-Firm Resources and Sustained Competitive Advantage. *management revue* **15**(1): 53-73

Dutta S, Narasimhan O, Rajiv S. 2005. Conceptualizing and measuring capabilities: methodology and empirical application. *Strategic Management Journal* **26**: 277-285

Dwyer FR, Tanner JF. 2002. *Business marketing: Connecting strategy, relationships and learning* (2 ed.). McGraw-Hill: Boston, MA

Dyer JH. 2000. *Collaborative Advantage: Winning through Extended Enterprise Supplier Networks*. Oxford University Press: Oxford

Dyer JH, Cho DS, Chu W. 1998a. Strategic Supplier Segmentation: The Next "Best Practice" in Supply Chain Management. *California Management Review* **40**(2): 57-77

Dyer JH, Hatch NW. 2004. Using Supplier Networks to Learn Faster. *MIT Sloan Management Review* **45**(3): 57-63

Dyer JH, Singh H. 1998b. The Relational View: Cooperative Strategy and Sources of Interorganizational Competitive Advantage. *Academy of Management Review* **23**(4): 660-679

Eberl M. 2004. Formative und reflektive Indikatoren im Forschungsprozess: Entscheidungsregeln und die Dominanz des reflektiven Modells., *Schriften zur Empirischen Forschung und Quantiativen Unternehmensplanung. Heft 19*. LMU München: München

Eger M. 1998. Stategischer Einkauf gibt Impulse für die Produktentwicklung, *Beschaffung Aktuell*: 38-42

Eilles A, Bartels M, Brunsman B. 2003. Managing the relationship portfolio. *Journal of Business Strategy* **24**(4): 30-34

Eisenhardt KM, Martin JA. 2000. Dynamic Capabilities: What Are They? *Strategic Management Journal* **21**: 1105-1121

Ellram LM. 1990. The supplier selection decision in strategic partnerships. *Journal of Purchasing and Materials Management* **26**(4): 8-14

Ellram LM, Baohong L. 2002a. The Financial Impact of Supply Management. *Supply Chain Management Review* **6**(6): 30-37

Ellram LM, Birou LM. 1995a. *Purchasing for Bottom Line Impact. Improving the Organization Through Strategic Procurement*. McGraw-Hill: New York, NY

Ellram LM, Carr AS. 1994. Strategic Purchasing: A History and Review of the Literature. *International Journal of Purchasing and Materials Management* **30**(2): 9-18

Ellram LM, Hendrick TE. 1995b. Partnering Characteristics: A Dyadic Perspective. *Journal of Business Logistics* **16**(1): 41-64

Ellram LM, Zsidisin GA, Siferd SP. 2002b. The Impact of Purchasing and Supply Management Activities on Corporate Success. *The Journal of Supply Chain Management: A Global Review of Purchasing and Supply* **38**(1): 4-17

Engel RJ. 2005. Role of Supply Chain Management in Revenue Enhancement, *90th Annual International Supply Management Conference*: San Diego, CA

Entchelmeier A, Jahns C. 2006. Leveraging Procurement Savings with Performance Measurement. European Case Clearing House: Cranfield

Ericksen PD. 2001. Managing the Extended Enterprise, *Purchasing Today*: 58

Eßig M. 1999. *Cooperative sourcing: Erklärung und Gestaltung horizontaler Beschaffungskooperationen in der Industrie*. Campus Verlag: Frankfurt am Main

Evans JR, Berman B. 2001. Conceptualizing and Operationalizing the Business-to-Business Value Chain. *Industrial Marketing Management* **30**: 135-148

Falk RF, Miller NB. 1992. *A Primer for Softmodelling*. University of Akron Press: Akron, OH

Farmer D. 1978. Developing Purchasing Strategies. *Journal of Purchasing and Materials Management* **14**(3): 6-11

Farrell D. 2005. Offshoring: Value Creation through Economic Change. *The Journal of Management Studies* **42**(3): 675-693

Fassott G. 2005. Die PLS-Pfadmodellierung: Entwicklungsrichtungen, Möglichkeiten, Grenzen. In F Bliemel, A Eggert, G Fassott, J Henseler (Eds.), *Handbuch PLS-Pfadmodellierung. Methode, Anwendung, Praxisbeispiele*: 19-29. Schäffer-Poeschel: Stuttgart

Ferguson WC, Hartley MF, Turner GB, Pierce EM. 1996. Purchasing's role in corporate strategic planning. *International Journal of Physical Distribution & Logistics Management* **26**(4): 51-62

Field A. 2005. *Discovering Statistics Using SPSS* (2 ed.). SAGE: London

Fine CH. 1998. *Clockspeed: Winning Industry Control in the Age of Temporary Advantage*. Perseus: Reading

Fisher ML. 1997. What is the Right Supply Chain for Your Product? *Harvard Business Review* **75**(2): 105-116

Fleisch E. 2001. *Das Netzwerkunternehmen. Strategien und Prozesse zur Steigerung der Wettbewerbsfähigkeit in the "Networked Economy"*. Springer: Berlin

Flint DJ, Woodruff RB. 2001. The Initiators of Changes in Customer's Desired Value. Results from a Theory Building Study. *Industrial Marketing Management* **30**: 321-337

Folker R. 1993. *Beschaffungsstrategien. Voraussetzungen, Methoden und EDV-Unterstützung einer problemadäquaten Auswahl*. Eul: Bergisch Gladbach

Fornell C. 1987. A Second Generation of Multivariate Analysis: Classification of Methods and Implications for Marketing Research. In MJ Houston (Ed.), *Review of Marketing*. American Marketing Association: Chicago, IL

Fornell C, Bookstein FL. 1982. Two structural equation models: LISREL and PLS applied to consumer exit-voice theory. *Journal of Marketing Research* **19**: 440-452

Fornell C, Cha J. 1994. Partial Least Squares. In RP Bagozzi (Ed.), *Advanced Methods of Marketing Research*: 52-78. Blackwell: Cambridge

Fornell C, Larcker DF. 1981. Evaluating Structural Equation Models with Unobservable Variables and Measurement Errors. *Journal of Marketing Research* **18**(2): 39-50

Foss NJ. 1999. Networks, capabilities, and competitive advantage. *Scandinvian Journal of Management* **15**(1): 1-15

Fraser P, Farrukh C, Gregory M. 2003. Managing product development collaborations - a process maturity approach. *Proceedings of the Institution of Mechanical Engineers. Part B. Engineering Manufacture* **21**(11): 1499-1519

Fredriksson P, Gadde L-E. 2003. Evaluation of Supplier Performance. The case of Volvo Car Corporation and its module suppliers, *19th Annual IMP Conference*: Lugano

Freeman VT, Cavinato JL. 1990. Fitting Purchasing to the Strategic Firm: Frameworks, Processes, and Values. *Journal of Purchasing and Materials Management* **26**(4): 6-10

Fritz W, Möllenberg A, Dees H. 2005. Erfolgsfaktoren von Internet-Auktionen: Eine empirische Analyse mit PLS. In F Bliemel, A Eggert, G Fassott, J Henseler (Eds.), *Handbuch PLS-Pfadmodellierung. Methode, Anwendung, Praxisbeispiele*: 255-274. Schäffer-Poeschel: Stuttgart

Fröhlich-Glantschnig E. 1997. Der Entscheidungsprozess wird transparent, *Beschaffung Aktuell*: 32-34

Fuchs PH, Mifflin KE, Miller D, Whitney JO. 2000. Strategic Integration. Competing in the Age of Capabilities. *California Management Review* **42**(3): 118-147

Fuhry G, Reinecke N. 2003. Einkauf von Innovationen. Selbst erfinden oder anderswo erfolgreich entdecken, *Beschaffung Aktuell*: 26-30

Gadde L-E, Jellbo O. 2002. System sourcing - opportunities and problems. *European Journal of Purchasing & Supply Management* **8**: 43-51

Gaffinen PZ, Holliger R. 1994. Supply Management bei ABB, *Thexis*: 29-37

Gallear D, Ghobadian A. 2004. Study of supply-chain purchasing strategy, *Journal of the Institute of Logistics and Transport*: 52-54

Gallery C, Hughes J. 2002. Linking procurement to corporate strategy: new goals for procurement - a value based strategy, *Insight Guide 10*. future purchasing: Surrey

Gälweiler A. 1986. *Unternehmensplanung. Grundlagen und Praxis*. Campus Verlag: Frankfurt am Main

Gälweiler A. 1987. *Strategische Unternehmensführung*. Campus Verlag: Frankfurt am Main

Ganesan S. 1994. Determinants of long-term orientation in buyer-seller relationships. *Journal of Marketing* **58**: 1-19

Gassmann O. 2003. Lieferanten als Innovationsquelle in F&E. In R Boutellier, SM Wagner, HP Wehrli (Eds.), *Handbuch Beschaffung. Strategien, Methoden, Umsetzung*: 631-646. Hanser: München

Gefen D, Straub DW. 2005. A practical guide to factorial validity using PLS-Graph: Tutorial and annotated example. *Communications of the Association for Information Systems* **16**: 91-109

Gefen D, Straub DW, Boudreau M-C. 2000. Structural Equation Modeling and Regression: Guidelines for Research Practice. *Communications of the Association for Information Systems* **4**(7): 2-76

Geisser S. 1975. A Predictive Approach to the Random Effect Model. *Biometrika* **61**(1): 101-107

Gemünden HG, Ritter T, Heydebreck P. 1996. Network Configuration and Innovation Success: An Empirical Analysis in German High-Tech Industries. *International Journal of Research in Marketing* **13**: 449-462

Gerpott TJ, Jakopin NM. 2005. International Marketing Standardization and Financial Performance of Mobile Network Operators - An Empirical Analysis. *Schmalenbach Business Review* **57**(3): 198-228

Giddens A. 1984. *The constitution of society*. Polity Press: Cambridge

Gilbert X, Strebel P. 1987. Strategies to outpace the competition. *Journal of Business Strategy* **8**(1): 28-36

Gilmore JH, Pine JB. 1997. The Four Faces of Mass Customization. *Harvard Business Review* **75**(1): 91-101

Giunipero LC, Vogt J. 1997. Empowering the Purchasing Function: Moving to Team Decisions. *International Journal of Purchasing and Materials Management* **33**(1): 8-15

Glantschnig E. 1994. *Merkmalsgestützte Lieferantenbewertung*. Fördergesellschaft Produkt-Marketing e.V.: Köln

Gleich R. 2001. *Das System des Performance Measurement. Theoretisches Grundkonzept, Entwicklungs- und Anwendungsstand*. Vahlen: München

Gleich R. 2005. Innovative Netzwerksteuerung in der produzierenden Industrie. In R Gleich (Ed.), *Managementherausforderungen in der produzierenden Industrie*: 1-16. Verlag Wissenschaft & Praxis: Sternenfels

Gluck FW, Kaufman SP, Walleck S. 1980. Strategic Management for Competitive Advantage. *Harvard Business Review* **58**(4): 154-161

Goebel DJ, Marshall GW, Locander WB. 2003. Enhancing Purchasing's Strategic Reputation: Evidence and Recommendations for Future Research. *Journal of Supply Chain Management* **39**(2): 4-14

Goffin K, Szwejczewski M, New C. 1997. Managing suppliers: when fewer can mean more. *International Journal of Physical Distribution & Logistics Management* **27**(7): 422-436

Goh M, Lau GT, Neo L. 1999. Strategic Role and Contribution of Purchasing in Singapore: A Survey of CEOs. *The Journal of Supply Chain Management* **35**(4): 12-22

Gomez P, Probst G. 1997. *Die Praxis des ganzheitlichen Problemlösens*. Haupt: Bern

González ME, Quesada G, Monge CAM. 2004. Determining the importance of the supplier selection process in manufacturing: a case study. *International Journal of Physical Distribution & Logistics Management* **34**(6): 492-504

Göpfert I. 1999. Stand und Entwicklung der Logistik. Herausbildung einer betriebswirtschaftlichen Teildisziplin. *Logistik Management* **1**(1): 19-33

Gottfredson M, Puryear R, Phillips S. 2005. Strategic Sourcing: From Periphery to the Core. *Harvard Business Review* **83**(2): 132-139

Götz O, Liehr-Gobbers K. 2004. Analyse von Strukturgleichungsmodellen mit Hilfe der Partial-Least-Squares (PLS)-Methode. *Die Betriebswirtschaft* **64**(6): 714-738

Green RF, Lisboa J, Yasin MM. 1993. Porter's (1980) Generic Strategies in Portugal. *European Business Review* **93**(2): 3-10

Gregory RE. 1986. Source selection: A matrix approach. *Journal of Purchasing and Materials Management* **22**(2): 24-29

Grimes JL. 2004. Sarbanes-Oxley and Suppliers, *TheMeetingProfessional*, Vol. 12: 152-153

Grundlach G, Achrol R. 1993. Governance in Exchange: Contract Law and Its Alternatives. *Journal of Public Policy & Marketing* **12**(4): 141-155

Guinipero LC, Brewer DJ. 1993. Performance Based Supplier Selection Systems under Total Quality Management. *International Journal of Purchasing and Materials Management* **29**(4): 35-41

Guinipero LC, Eltanatwy RA. 2004a. Securing the upstream supply chain: a risk management approach. *International Journal of Physical Distribution & Logistics Management* **34**(9): 698-713

Guinipero LC, Handfield RB. 2004b. *Purchasing Education and Training II*. CAPS: Tempe, AZ

Haenlein M, Kaplan AM. 2004. A Beginner's Guide to Partial Least Squares Analysis. *Understanding Statistics* **3**(4): 283-297

Hagel J, Singer M. 1999. *Net Worth: shaping markets when customers make the rules*. Harvard Business School Press: Boston, MA

Hair J, Anderson RE, Tatham RL, Black WC. 1998. *Multivariate Data Analysis* (5 ed.). Englewood Cliffs: New Jersey

Hakansson H (Ed.). 1982. *International Marketing and Purchasing of Industrial Goods: An Interaction Approach*. Wiley: Chichester

Halldorsson A, Skjott-Larsen T, Kotzab H. 2003. Interorganizational Theories Behind Supply Chain Management - Discussion and Applications. In S Seuring, M Müller, M Goldbach (Eds.), *Strategy and Organization in Supply Chains*: 31-45. Physica-Verlag: Heidelberg

Hamm V. 1998. Zukunftsperspektiven des Einkaufs, *Beschaffung Aktuell*: 32-35

Handfield RB, Nichols EL. 2004. Key issues in global supply base management. *Industrial Marketing Management* **33**: 29-35

Hansmann K-W, Ringle M. 2004. SmartPLS Benutzerhandbuch 1.01. University of Hamburg: Hamburg

Harland CM. 1996. Supply chain management: relationships, chains and networks. *British Journal of Management* 7(Special Issue): 163-180

Harland CM, Brenchley R, Walker H. 2003. Risk in supply networks. *Journal of Purchasing & Supply Management* **9**: 51-62

Harland CM, Lamming R, Cousins PD. 1999. Developing the concept of supply strategy. *International Journal of Operations & Production Management* **19**(7): 650-673

Harting D. 1995. TQM - neueste Modewelle oder strategischer Wettbewerbsfaktor, *Beschaffung Akutell*: 30-31

Hartmann E, Moser R. 2005. *Supply Techniques*. Verlag Wissenschaft & Praxis: Sternenfels

Hartmann E, Ritter T, Gemünden HG. 2001. Determining the Purchase Situation: Cornerstone of Supplier Relationship Management, *17th Annual IMP Conference*: Oslo

Hauser RM, Goldberger AS. 1971. The Treatment of Unobservable Variables in Path Analysis. In HL Costner (Ed.), *Sociological Methodology*: 81-117. SAGE: London

Heide JB, Stump RL. 1995. Performance Implications of Buyer-Supplier Relationships in Industrial Markets. A Transaction Cost Explanation. *Journal of Business Research* **32**: 57-66

Heuermann C. 2002. Internationalisierung und Logistikstrategie - Wettbewerbsstrategische Implikationen für die Gestaltung internationaler Logistiksysteme. In W Delfmann (Ed.), *Arbeitsbericht 104*. Universität Köln: Köln

Hewitt K, Money BR, Sharma S. 2002. An exploration of the moderating role of buyer corporate culture in industrial buyer-seller relationships. *Journal of the Academy of Marketing Science* **30**(3): 407-419

Hildebrand H, Koppelmann U. 2000. *Beziehungsmanagement mit Lieferanten. Konzepte, Instrument, Erfolgsnachweise*. Schäffer-Poeschel: Stuttgart

Himpel F. 1999. *Industrielle Beschaffungsnetzwerke. Theoretische Fundierung, Entwicklungsprinzipien und Gestaltungsaspekte*. Gabler: Wiesbaden

Hines P. 1996. Network sourcing. A discussion of causality within the buyer-supplier relationship. *European Journal of Purchasing & Supply Management* **2**(1): 7-20

Hines P, Lamming R, Jones DT, Cousins PD, Rich N. 2000. *Value Stream Management. Strategy and Excellence in the Supply Chain*. Prentice Hall: Harlow

Hitt MA, Ireland RD, Hoskisson RE. 1998. *Strategic Management. Competitiveness and Globalization* (4 ed.). South-Western College: Cincinnati

Hogan JE. 2001. Expected Relationship Value. A Construct, a Methodology for Measurement, and a Modeling Technique. *Industrial Marketing Management* **30**: 339-351

Hogan JE, Armstrong G. 2001. Toward a resource-based theory of business exchange relationships: The role of relational asset value. *Journal of Business-to-Business Marketing* **8**(4): 3-28

Hollinger R. 1992. Supply Management. In R Fickert, K Hässig (Eds.), *Megatrends*: 35-47. SVME: Aarau

Homburg C, Hildebrandt L. 1998. Die Kausalanalyse: Bestandsaufnahme, Entwicklungsrichtungen, Problemfelder. In L Hildebrandt, C Homburg (Eds.), *Die Kausalanalyse: ein Instrument der betriebswirtschaftlichen Forschung*: 15-43. Schäffer-Poeschel: Stuttgart

Hoopes DG, Madsen TL, Walker G. 2003. Why is there a resource-based view? Toward a theory of competitive heterogeneity. *Strategic Management Journal* **24**: 889-902

Houghton T, Markham B, Tevelson B. 2002. Thinking strategically about supply management. *Supply Chain Management Review* **6**(5): 32-38

Howard T, Pollock T. 1999. From I-O Economics' S-C-P Paradigm Through Strategic Groups to Competence-Based Competition: Reflections on the Puzzle of Competitive Strategy. *British Journal of Management* **10**: 127-140

Hoyt J, Hug F. 2000. From arms-length to collaborative relationships in the supply chain. An evolutionary process. *International Journal of Physical Distribution & Logistics Management* **30**(9): 750-764

Hughes J. 2005. Supplier metrics that matter, *CPO Agenda*, Vol. 1: 19-23

Hugos M. 2003. *Essentials of Supply Chain Management*. Wiley: Hoboken, NJ

Hulland J. 1999. Use of Partial Least Squares (PLS) in Strategic Management Research: A Review of Four Recent Studies. *Strategic Management Journal* **20**: 195-204

Humphreys PK, Li A, Chan P. 2004. The impact of supplier development on buyer-supplier performance. *Omega* **32**(2): 131-143

Humphreys PK, McIvor R, McAleer W. 2000. Re-engineering the purchasing function. *European Journal of Purchasing & Supply Management* **6**: 85-93

Hutchins G. 2002. *Supply Management Strategies*. QPE: Portland, OR

Injazz JC, Paulraj A, Lado AA. 2004. Strategic purchasing, supply management, and firm performance. *Journal of Operations Management* **22**: 505-523

Jaccard J, Wan CK. 1996. *LISREL approaches to interaction effects in multiple regression*. SAGE: Thousand Oaks, CA

Jackson B. 1985. *Winning and Keeping Industrial Customers*. Lexington Books: Lexington, MA

Jacobson R. 1992. The 'Austrian' School of strategy. *Academy of Management Review* **17**: 782-807

Jagpal HS. 1982. Multicollinearity in Structural Equation Models With Unobservable Variables. *Journal of Marketing Research* **14**: 431-439

Jahns C. 1999. *Integriertes Strategisches Management. Neue Perspektiven zur Theorie und Praxis des strategischen Managements*. Verlag Wissenschaft & Praxis: Sternenfels/Dissertation TU München 1998

Jahns C. 2003a. *Arbeitsbuch Managementtheorie*. Verlag Wissenschaft & Praxis: Sternenfels

Jahns C. 2003b. *Arbeitsbuch Strategisches Management*. Verlag Wissenschaft & Praxis: Sternenfels

Jahns C. 2005a. Supply Controlling. *Controlling. Zeitschrift für erfolgsorientierte Unternehmenssteuerung* **17**(6): 349-358

Jahns C. 2005b. *Supply Management. Neue Perspektiven eines Managementansatzes für Einkauf und Supply*. Verlag Wissenschaft & Praxis: Sternenfels/Habilitation TU München

Jahns C. 2006. Excellence in Purchasing and Supply Management, *European Leaders Executive Network - PSM*: London

Jahns C, Hartmann E, Wagner A. 2005a. *Marketing Spend Survey 2005*. Verlag Wissenschaft & Praxis: Sternenfels

Jahns C, Henke M. 2005b. *Arbeitsbuch Supply Risk Management*. Verlag Wissenschaft & Praxis: Sternenfels

Jahns C, Moser R. 2005c. *Supplier Value - How to Define and Integrate the Value of Your Suppliers*. Verlag Wissenschaft & Praxis: Sternenfels

Jahns C, Moser R, Hartmann E. 2006. *Schaffung von Wettbewerbsvorteilen durch die Auswahl strategischer Lieferanten*. Verlag Wissenschaft & Praxis: Sternenfels

Janker CG. 2004. *Multivariate Lieferantenbewertung. Empirisch gestützte Konzeption eines anforderungsgerechten Bewertungssystems*. DUV: Wiesbaden

Jap SD. 1999. Pie-expansion efforts: Collaboration processes in buyer-supplier relationships. *Journal of Marketing Research* **36**: 461-475

Jap SD. 2001. Perspectives on joint competitive advantages in buyer-supplier relationships. *International Journal of Research in Marketing* **18**: 19-35

Jarvis CB, MacKenzie SB, Podsakoff PM. 2003. A Critical Review of Construct Indicators and Measurement Model Misspecification in Marketing and Consumer Research. *Journal of Consumer Research* **30**: 199-218

Johansson JK, Yyp GS. 1994. Exploiting globalization potential: U.S. and Japanese strategies. *Strategic Management Journal* **15**: 579-601

Johnsen T, al e. 2000. Networking activities in supply networks. *Journal of Strategic Management* **8**: 161-181

Johnson FP, Leenders MR. 2004. *Supply's Organizational Roles and Responsibilities*. CAPS: Tempe, AZ

Johnston DA, McCutcheon DM, Stuart FI, Kerwood H. 2004. Effects of supplier trust on performance of cooperative supplier relationships. *Journal of Operations Management* **22**: 23-38

Jöreskog KG. 1970. A general method for analysis of covariance structures. *Biometrika* **57**: 239-251

Jöreskog KG, Sörbom D. 1996. *LISREL 8: User's reference guide*. Scientific Software International: Chicago, IL

Jorgensen H. 2005. Methods & Elements of a Solid Risk Management Strategy. *Risk Management* **52**(7): 53

Jüttner U. 2003. Risiko- und Krisenmanagement in Supply Chains. In R Boutellier, SM Wagner, HP Wehrli (Eds.), *Handbuch Beschaffung. Strategien, Methoden, Umsetzung*: 775-796. Hanser: München

Kannan VR, Tan KC. 2002. Supplier Selection and Assessment: Their Impact on Business Performance. *The Journal of Supply Chain Management: A Global Review of Purchasing and Supply* **38**(4): 11-21

Kannan VR, Tan KC. 2003. Attitudes of US and European managers to supplier selection and assessment and implications for business performance. *Benchmarking* **10**(5): 472-489

Kappeler B. 2004. Karl für Sparsame, *NZZ am Sonntag*: 27: Zürich

Karimi J, Somers TM, Gupta YP. 2004. Impact of Environmental Uncertainty and Task Characteristics on User Satisfaction with Data. *Information Systems Research* **15**(2): 175-193

Katsikeas CS, Paparoidamis NG, Katsikea E. 2004. Supply source selection criteria: The impact of supplier performance on distributor performance. *Industrial Marketing Management* **33**: 755-764

Katz R, Allen TJ. 1982. Investigating the not invented here (NIH) syndrome: A look at the performance, tenure, and communication patterns of 50 R&D project groups. *R&D Management* **12**: 7-19

Kauffman RG. 2002. Supply Management: What's in a Name? Or, Do We Know Who We Are? *Journal of Supply Chain Management* **38**(4): 46-50

Kaufman A, Wood CH, Theyel G. 2000. Collaboration and Technology Linkages: A Strategic Supplier Technology. *Strategic Management Journal* **21**: 649-663

Kaufmann L. 2001. *Internationales Beschaffungsmanagement. Gestaltung strategischer Gesamtsysteme und Management einzelner Transaktionen*. Gabler: Wiesbaden

Kaufmann L. 2004a. Supply Chain Portfolio - Ein Instrument für das situationsgerechte Management einzelner Kettenglieder. In H Baumgarten, H-P Wiendahl, J Zentes (Eds.), *Logistik-Management. Strategien - Konzepte - Praxisbeispiele*: 1-15. Springer: Berlin

Kaufmann L. 2004b. X-BSC - Measuring the Performance of Truly Strategic Supplier Relationships. *PRACTIX* **7**(3): 1-5

Kaufmann L, Germer T (Eds.). 2001. *Controlling internationaler Supply Chains. Positionierung - Instrumente - Perspektiven*. Lemmens Verlag: Bonn

Kaufmann L, Thiel C, Becker A. 2006. Innovative Beschaffungsportfolios und X-Balanced Scorecards - Komplementäre Instrumente für das strategische Controlling. *Controlling. Zeitschrift für erfolgsorientierte Unternehmenssteuerung* **18**(3): 125-132

Kemppainen K, Vepsälätinen APJ. 2003. Trends in industrial supply chains and networks. *International Journal of Physical Distribution & Logistics Management* **33**(8): 701-719

Kilgore MJ. 2004. Mitigating Supply Chain Risks, *89th Annual International Supply Management Conference*: Phoenix, AZ

Kinder T. 2003. Go with the flow - a conceptual framework for supply relations in the era of the extended enterprise. *Research Policy* **32**: 503-523

Kleinau AC. 1994. *Zur Strategie der Lieferantenentwicklung*. Lang: Frankfurt am Main

Kligge CG. 1992. *Strategische Beschaffung und Rückwärtsintegration*. Lang: Frankfurt am Main

Knapp TM, Bichler K. 2000. Beschaffung von Nicht-Produktionsmaterial, *Beschaffung Aktuell*: 42-46

Knudsen D. 2003. Aligning corporate strategy, procurement strategy and e-procurement tools. *International Journal of Physical Distribution & Logistics Management* **33**(8): 720-734

Koppelmann U. 2002a. Beschaffung als Forschungs- und Lehrzweig. In D Hahn, L Kaufmann (Eds.), *Handbuch industrielles Management*, 2 ed.: 947-957. Gabler: Wiesbaden

Koppelmann U. 2002b. *Beschaffungsmarketing* (4 ed.). Springer: Berlin

Kotabe M. 2002. Global Sourcing Strategy and Sustainability of Core Competences. Fundamental Differences between U.S. and Japanese Companies. In D Hahn, L Kaufmann (Eds.), *Handbuch industrielles Management*, 2 ed.: 241-262. Gabler: Wiesbaden

Kothandaraman P, Wilson DT. 2001. The Future of Competition. Value-Creating Networks. *Industrial Marketing Management* **30**: 379-389

Krafft M, Götz O, Liehr-Gobbers K. 2005. Die Validierung von Strukturgleichungsmodellen mit Hilfe des Partial-Least-Squares (PLS)-Ansatzes. In F Bliemel, A Eggert, G Fassott, J Henseler (Eds.), *Handbuch PLS-Pfadmodellierung. Methode, Anwendung, Praxisbeispiele*: 71-86. Schäffer-Poeschel: Stuttgart

Krapfel RE, Salmond D, Spekman R. 1991. A Strategic Approach to Managing Buyer-Seller Relationships. *European Journal of Marketing* **25**(9): 22-37

Krause DR, Handfield RB. 1999. *Developing a World-Class Supply Base*. CAPS: Tempe, AZ

Krause DR, Handfield RB, Scannell TV. 1998. An empirical investigation of supplier development: reactive and strategic processes. *Journal of Operations Management* **17**: 39-58

Krause DR, Pagell M, Curkovic S. 2001. Toward a measure of competitive priorities for purchasing. *Journal of Operations Management* **19**: 497-512

Krüger K. 2004. Negotiations in tender procedures: undermining best value for money? *Journal of Public Procurement* **4**(3): 397-437

Kühl S. 1995. *Wenn die Affen den Zoo regieren. Die Tücken der flachen Hierarchien* (2 ed.). Campus: Frankfurt am Main

Kwong CK. 2002. Combining scoring method and fuzzy expert systems approach to supplier asessment: a case study. *Integrated Manufacturing Systems* **17**(7): 512-519

Lambe JC, Spekman R. 1997. Alliances, external technology acquisition, and discontinuous technological change. *Journal of Product Innovation Management* **14**(2): 102-114

Lambert DM, Stock JR, Ellram LM. 1998. *Fundamentals of Logistics Management*. Irwin/McGraw-Hill: Boston, MA

Lamming R, Johnsen T, Zheng J, Harland CM. 2000. An initial classification of supply networks. *International Journal of Operations & Production Management* **20**(6): 675-691

Large R. 2000. *Strategisches Beschaffungsmanagement. Eine praxisorientierte Einführung* (2 ed.). Gabler: Wiesbaden

Laseter TM. 1998. *Balanced Sourcing. Cooperation and Competition in Supplier Relationships*. Wiley: San Francisco, CA

Lee H. 2000. Creating Value through Supply Chain Integration. *Supply Chain Management Review* **4**(5): 30-49

Lee H. 2002. Aligning supply chain strategies with product uncertainties. *California Management Review* **44**(3): 105-119

Leek S. 2003. Interactions, relationships and networks in a changing world. *Industrial Marketing Management* **32**: 87-90

Leenders MR, Fearon HE, Flynn AE, Johnson FP. 2002. *Purchasing and Supply Management* (12 ed.). McGraw-Hill: New York, NY

Lemke F, Goffin K, Szwejczewski M, New C. 2000. Mit leistungsfähigen Gleichgesinnten verbünden, *Beschaffung Aktuell*: 34-36

Leverick F, Cooper R. 1998. Partnerships in the Motor Industry: Opportunities and Risks for Suppliers. *Long Range Planning* **31**(1): 72-81

Lieberman MB, Montgomery DB. 1988. First-Mover Advantages. *Strategic Management Journal* **9**(5): 41-58

Linder JC, Jarvenpaa S, Davenport TH. 2003. Toward an Innovation Sourcing Strategy. *MIT Sloan Management Review 2003* **44**(3): 43-49

Little N. 2004. Price, Cost, Value, Benefit, Gain, Worth...Which Way Do I Turn?, *89th Annual International Supply Management Conference*: Phoenix, AZ

Little RJA. 1988. A Test of Missing Completely at Random for Multivariate Data with Missing Values. *Journal of the American Statistical Association* **83**(404): 1198-1202

Loasby BJ. 1998. The Organization of Capabilities. *Journal of Economic Behaviour and Organization* **35**: 139-160

Locke D. 1996. *Global Supply Management. A Guide to International Purchasing*. McGraw-Hill: Boston, MA

Locke R, Anklesaria J. 1994. Selection of Currency And Hedging Strategy in Global Supply Management, *79th Annual International Purchasing Conference*: Atlanta, GA

Lohmöller J-B. 1989. *Latent variable path modelling with partial least squares*. Physica-Verlag: Heidelberg

Lowson RH. 2003. The nature of an operation strategy: combining strategic decisions from the resource-based and market-driven viewpoints. *Management Decision* **41**(6): 538-549

Luhmann N. 1996. *Soziale Systeme. Grundriß einer allgemeinen Theorie* (6 ed.). Suhrkamp: Frankfurt am Main

Lynskey MJ. 1999. The Transfer of Resources and Competencies for Developing Technological Capabilities - The Case of Fujitsu-ICL. *Technology Analysis & Strategic Management* **11**(3): 317-336

Lysons K, Farrington B. 2006. *Purchasing and Supply Chain Management* (7 ed.). Prentice Hall: Harlow

Magretta J. 1998. The power of Virtual Integration: An Interview With Dell Computer's Michael Dell. *Harvard Business Review* **76**(2): 72-84

Marino KE. 1996. Developing consensus on firm competencies and capabilities. *Academy of Management Executive* **10**(3): 40-51

Markowitz H. 1952. Portfolio Selection. *The Journal of Finance* **7**(1): 77-91

Matthews PA, al e. 2000. The Future of Purchasing and Supply. In JL Cavinato, RG Kauffman (Eds.), *The Purchasing Handbook. A Guide for the Purchasing and Supply Professional*, 6 ed.: 3-33. McGraw-Hill: New York, NY

McGinnis MA, Vallopra RM. 1998. *Purchasing and supplier involvement: new product development and production/operations process development and improvement*. CAPS: Tempe, AZ

McGinnis MA, Vallopra RM. 1999. Purchasing and Supplier Involvement in Process Improvement: A Source of Competitive Advantage. *Journal of Supply Chain Management* **35**(4): 42-50

McIvor R, Humphreys PK. 2004. Early supplier involvement in the design process: lessons from the electronics industry. *Omega* **32**: 179-199

Meier MC. 1988. *Strategisches Materialressourcen-Management*. SVME: Aarau

Meulbroek L. 2002. The promise and challenge of integrated risk management. *Risk Management and Insurance Review* **5**(1): 55-66

Meyer C. 1990. *Beschaffungsziele* (2 ed.). Fördergesellschaft Produkt-Marketing e.V.: Düsseldorf

Michiels WL. 2000. Biz, SC strategies should be linked, *Electronic Buyer News*: 82-84

Michiels WL, Eckardt TJ. 2001. Strategic Purchasing: Not for Near-Sighted People, *NAPM 86th Annual International Puchasing Conference*: Phoenix, AZ

Miller D, Dess GG. 1993. Assessing Porter's (1980) model in terms of its generalizability, accuracy and simplicity. *Journal of Management Studies* **30**(4): 553-585

Milligan B. 1999. Purchasing pros stumble toward strategic supply management, *Purchasing*: 20-21

Mintzberg H, Walters JA. 1985. Of strategies, deliberate and emergent. *Strategic Management Journal* **6**: 257-272

Mitchell P. 2002. Integrating Strategic Sourcing and Supply Management. *Supply Chain Management Review* **6**(5): 15-16

Moharty RP, Deshmukh SG. 1993. Use of analytic hierarchic process for evaluating sources of supply. *International Journal of Physical Distribution and Logistics Management* **23**(3): 22-28

Möhrstädt DG, Bogner P, Paxian S. 2001. *Electronic Procurement planen, einführen, nutzen*. Schäffer-Poeschel: Stuttgart

Mol MJ. 2003. Purchasing's strategic relevance. *Journal of Purchasing & Supply Management* **9**(1): 43-50

Möller KEK, Törrönen P. 2003. Business supplier's value creation potential. A capability-based analysis. *Industrial Marketing Management* **32**: 109-118

Monczka RM, Callahan TJ, Nichols EL. 1995. Predictors of relationships among buying and supplying firms. *International Journal of Physical Distribution & Logistics Management* **25**(10): 45-59

Monczka RM, Handfield RB, Scannell TV, Ragatz GL, Frayer DJ. 2000a. *New Product Development. Strategies for Supplier Integration*. Amer Society for Quality: Milwaukee, WI

Monczka RM, Morgan JP. 1994. Today's measurements just don't make it!, *Purchasing*: 30-33

Monczka RM, Morgan JP. 1996. Supplier integration. A new level of supply chain management, *Purchasing*: 110-113

Monczka RM, Morgan JP. 2000b. Competitive Supply Strategies for the 21. Century, *Purchasing*: 48-59

Monczka RM, Trent RJ, Handfield RB. 2002. *Purchasing and Supply Chain Management*. South Western: Cincinnati, OH

Moody PE. 2001. Strategic Purchasing Remains an Oxymoron. *MIT Sloan Management Review* **42**(2): 18

Morash EA. 2001. Supply Chain Strategies, Capabilities, and Performance. *Transportation Journal* **41**(1): 37-54

Moser R, Jahns C. 2005. Supplier Evaluation at EADS. In European Case Clearing House (Ed.): Cranfield

Müller M. 2003. The Use of Information Technologies in Supply Chains - A Transaction Cost Analysis. In S Seuring, M Müller, M Goldbach (Eds.), *Strategy and Organization in Supply Chains*: 18-30. Physica-Verlag: Heidelberg

Müller-Stewens G, Lechner C. 2003. *Strategisches Management. Wie strategische Initiativen zum Wandel führen* (2 ed.). Schäffer-Poeschel: Stuttgart

Narasimhan R, Carter JR. 1998. Linking Business Unit and Material Sourcing Strategies. *Journal of Business Logistics* **19**(2): 155-171

Narasimhan R, Das A. 2001a. The impact of purchasing integration and practices on manufacturing performance. *Journal of Operations Management* **19**: 593-609

Narasimhan R, Talluri S, Mendez D. 2001b. Supplier Evaluation and Rationalization via Data Envelopment Analysis: An Empirical Examination. *The Journal of Supply Chain Management: A Global Review of Purchasing and Supply* **37**(3): 28-37

Narver JC, Slater SF. 1990. The Effect of a Market Orientation on Business Profitability. *Journal of Marketing* **54**(4): 20-35

Nicolai A. 2004. Der "trade-off" zwischen "rigour" und "relevance" und seine Konsequenzen für die Managementwissenschaften. *Zeitschrift für Betriebswirtschaft* **56**: 99-118

Nollet J, Ponce S, Campbell M. 2005. About "strategy" and "strategies" in supply management. *Journal of Purchasing & Supply Management* **11**(4): 129-140

Noonan R, Wold H. 1982. PLS path modeling with indirectly observed variables: A comparison of alternative estimates for latent variables. In KG Jöreskog, H Wold (Eds.), *Systems under indirect observations: Causality, structure, prediction*: 75-94. North Holland: Amsterdam

Normann R, Ramirez R. 1993. From Value Chain to Value Constellation: Designing Interactive Strategy. *Harvard Business Review* **71**(4): 65-77

Norrman A, Jansson U. 2004. Ericsson's proactive supply chain risk management approach after a serious sub-supplier accident. *International Journal of Physical Distribution & Logistics Management* **34**(5): 434-456

Nunnally JC, Bernstein IH. 1994. *Psychometric Theory* (3 ed.). McGraw-Hill: New York, NY

Odgen JA, Petersen KJ, Carter J, Monczka RM. 2005. Supply Management Strategies for the Future: A Delphi Study. *The Journal of Supply Chain Management: A Global Review of Purchasing and Supply* **41**(3): 29-48

Okamuro H. 2001. Risk sharing in the supplier relationship: new evidence from the Japanese automotive industry. *Journal of Economic Behavior & Organization* **45**: 361-381

Olsen RF, Ellram LM. 1997. A Portfolio Approach to Supplier Relationships. *Industrial Marketing Management* **26**: 101-113

Ortiz de Guinea A, Kelley H, Hunter GM. 2005. Information Systems Effectiveness in Small Businesses: Extending a Singaporean Model in Canada. *Journal of Global Information Management* **13**(3): 55-79

Östring P. 2003. *Profit-Focused Supplier Management. How to Identify Risks and Recognize Opportunities*. American Management Association: New York, NY

Owens SC, Coe BJ. 1994. Sourcing Preference Scale: Measuring Preferences of Purchasing Managers for Single Sourcing or Multiple Sourcing of Products. *Industrial Marketing Management* **23**: 171-180

Pagell M. 2004. Understanding the factors that enable and inhibit the integration of operations, purchasing and logistics. *Journal of Operations Management* **22**: 459-487

Pagell M, Krause DR. 2002. Strategic consensus in the internal supply chain: exploring the manufacturing-purchasing link. *International Journal of Production Research* **40**(13): 3075-3092

Patterson JL. 1996. An Investigating of Conflict Resolution between Buyers and Suppliers in Strategic Long-Term Cooperative Relationships., *NAPM 81st Annual International Purchasing Conference*: Chicago, IL

Paulraj A, Chen IJ. 2005. Driving forces of strategic supply management: a preliminary empirical investigation. *International Journal of Integrated Supply Management* **1**(3): 312-334

Payne A, Holt S. 2001. Diagnosing Customer Value: Integrating the Value Process and Relationship Marketing. *British Journal of Management* **12**: 159-182

Pearson JN, Ellram LM. 1995. Supplier Selection and Evaluation in Small versus Large Electronics Firms. *Journal of Small Business Management* **33**(3): 53-65

Pearson JN, Gritzmacher KJ. 1990. Integrating Purchasing into Strategic Management. *Long Range Planning* **23**(3): 91-99

Penrose E. 1959. *The Theory of the Growth of the Firm*. Oxford University Press: Oxford

Peter M. 1996. *Early Supplier Involvement (ESI) in Product Development*. Difo-Druck: Bamberg/Dissertation University of St. Gallen

Peteraf M. 1993. The cornerstones of competitive advantage: a resource based view. *Strategic Management Journal* **14**: 179-191

Petroni A, Braglia M. 2000. Vendor Selection Using Principal Component Analysis. *The Journal of Supply Chain Management: A Global Review of Purchasing and Supply* **36**(2): 63-72

Pfisterer J. 1988. *Beschaffungskontrolle*. Fördergesellschaft Produkt-Marketing e.V.: Düsseldorf

Picot A. 1982. Transaktionskostenansatz in der Organisationstheorie: Stand der Diskussion und Aussagewert. *Die Betriebswirtschaft* **42**(2): 267-279

Picot A. 1991. Ein neuer Ansatz zur Gestaltung der Leistungstiefe. *Zeitschrift für betriebswirtschaftliche Forschung* **43**: 336-357

Picot A, Dietl H, Franck E. 1997. *Organisation: eine ökonomische Perspektive*. Schäffer-Poeschel: Stuttgart

Picot A, Reichwald R, Wigand RT. 2003. *Die grenzenlose Unternehmung. Information, Organisation und Management* (5 ed.). Gabler: Wiesbaden

Piller FT. 2001. *Mass Customization. Ein Wettbewerbskonzept für das Informationszeitalter* (2 ed.). Gabler: Wiesbaden

Pine BJ. 1993. *Mass Customization. The New Frontier in Business Competition*. Harvard Business School Press: Boston, MA

Piontek J. 1999. *Beschaffungscontrolling* (2 ed.). Oldenbourg Verlag: München

Pitt M, Clarke K. 1999. Competing on Competence: A Knowledge Perspective on the Management of Strategic Innovation. *Technology Analysis & Strategic Management* **11**(3): 301-316

Porter ME. 1980. *Competitive Strategy: Techniques for Analyzing Industries and Competitors*. Free Press: New York, NY

Porter ME. 1985. *Competitive Advantage: Creating and Sustaining Superior Performance*. Free Press: New York, NY

Porth SJ. 2003. *Strategic Management. A Cross-Functional Approach*. Prentice Hall: Upper Saddle River, NJ

Powell TC. 2001. Competitive advantage: logical and philosophical considerations. *Strategic Management Journal* **22**: 875-888

Powell WW, Koput KW, Smith-Doerr L. 1996. Interorganizational collaboration and the locus of innovation: Networks of learning in biotechnology. *Administrative Science Quaterly* **41**(1): 116-145

Prahalad CK, Hamel G. 1990. The core competency of the corporation. *Harvard Business Review* **68**(3): 79-91

Priem RL, Butler JE. 2001. Is the resource-based "view" a useful perspective for strategic management reseach? *Academy of Management Review* **26**(1): 22-40

Procurement Strategy Council. 2002. Supplier Relationship and Segmentation Models, *Corporate Executive Board*: 1-20: Washington

Quayle M. 2001. Purchasing in the UK and Switzerland: an empirical study of sourcing decisions. *European Business Review* **13**(1): 42-59

Ragatz GL, Handfield RB, Petersen KJ. 2002. Benefits associated with supplier integration into new product development under conditions of technology uncertainty. *Journal of Business Research* **55**(5): 389-400

Rajagopal S, Bernard KN. 1993. Strategic procurement and competitive advantage. *Journal of Purchasing and Materials Management* **29**(4): 13-20

Ramsay J. 2001a. Purchasing's strategic irrelevance. *European Journal of Purchasing & Supply Management* **7**: 257-263

Ramsay J. 2001b. The Resource Based Perspective, Rents, and Purchasing's Contribution to Sustainable Competitive Advantage. *Journal of Supply Chain Management* **37**(3): 38-47

Ravald A, Grönroos C. 1996. The Value Concept and Relationship Marketing. *European Journal of Marketing* **30**: 19-30

Rayport JF, Sviokla JJ. 1999. Managing the Marketplace. *Harvard Business Review* **77**(6): 141-150

Reck RF, Long BG. 1988. Purchasing: A Competitive Weapon. *Journal of Purchasing and Materials Management* **24**(3): 2-8

Reeves CA, Bednar DA. 1994. Defining Quality: Alternatives and Implications. *Academy of Management Review* **19**(3): 419-445

Reichwald R, Möslein K. 2000. Nutzenpotenziale und Nutzenrealisierung in verteilten Organisationsstrukturen. Experimente, Erprobungen und Erfahrungen auf dem Weg zur virtuellen Unternehmung. *Zeitschrift für Betriebswirtschaft* **70**(Ergänzungsheft 2): 117-136

Reinartz W, Krafft M, Hoyer WD. 2004. The Customer Relationship Management Process: Its Measurement and Impact on Performance. *Journal of Marketing Research* **41**: 293-305

Rice JB, Hoppe RM. 2001. SC vs. SC. *Supply Chain Management Review* **5**(5): 47-54

Riemer K, Klein S. 2002. Supplier Relationship Management. In K Hildebrand (Ed.), *Supplier Relationship Management*: 5-22. Dpunkt Verlag: Heidelberg

Riffner B, Weidelich R. 2001. *Professionelles Lieferantenmanagement. So arbeiten Kunden und Lieferanten erfolgreich zusammen.* Deutscher Wirtschaftsdienst: Köln

Ringle MC. 2004. Messung von Kausalmodellen. Ein Methodenvergleich. Arbeitspapier Nr. 14. Institut für Industriebetriebslehre und Organisation. University of Hamburg: Hamburg

Rink DR, Fox HW. 2003. Using the Product Life Cycle Concept to Formulate Actionable Purchasing Strategies. *Singapore Management Review* **25**(2): 73-89

Rondinelli D, Rosen B, Drori I. 2001. The Struggle for Strategic Alignment in Multinational Corporations: Managing Readjustment During Global Expansion. *European Management Journal* **19**(4): 404-416

Rossiter JR. 2002. The C-OAR-SE Procedure for Scale Development in Marketing. *International Journal of Research in Marketing* **19**(4): 305-335

Rozemeijer F. 2000. Creating Corporate Advantage in Purchasing. Technische Universität Eindhoven: Eindhoven

Rüegg-Stürm J. 2005. *The new St. Gallen Management Model - Basic Categories of an Approach to Integrated Management*. Palgrave: Houndmills, NY

Saaty TL. 1990. How to make a decision: The Analytic Hierarchy Process. *European Journal of Operational Research* **48**: 9-26

Sanchez R, Heene A. 1997. Reinventing Strategic Management: New Theory and Practice for Competence-based Competition. *European Management Journal* **15**(3): 303-317

Sanchez-Franco MJ, Roldan JL. 2005. Web acceptance and usage model - A comparison between goal-directed and experiental web users. *Internet Research* **15**(1): 21-48

Sanders DR, Manfredo MR. 2002. The Role of Value-at-Risk in Purchasing: An Application to the Foodservice Industry. *The Journal of Supply Chain Management: A Global Review of Purchasing and Supply* **38**(2): 38-45

Sarkar MB, Echambadi R, Cavusgil ST, Aulakh PS. 2001a. The Influence of Complementarity, Compatibility, and Relationship Capital on Alliance performance. *Academy of Marketing Science* **29**(4): 358-373

Sarkar MB, Echambadi R, Harrison JS. 2001b. Alliance Entrepreneurship and Firm Market Performance. *Strategic Management Journal* **22**: 701-711

Sarkis J, Talluri S. 2002. A Model for Strategic Supplier Selection. *The Journal of Supply Chain Management: A Global Review of Purchasing and Supply* **38**(1): 18-28

Sattler T, Milling P. 2005. Value Creation through Supplier Management and Supplier Involvement - A Conceptual Approach, *Budapest*: 1799-1808: 12th International EurOMA Conference

Scheuing EE. 1998. *Value-Added Purchasing*. Crisp Publications: Menlo Parc, CA

Schmitz J, Platts KW. 2003. Roles of supplier performance measurement: indication from a study in the automotive industry. *Management Decision* **41**(8): 711-721

Schmitz J, Platts KW. 2004. Supplier logistics performance measurement: Indications from a study in the automotive industry. *International Journal of Production Economics* **89**: 231-243

Schnell R, Hill PB, Esser E. 1992. *Methoden der empirischen Sozialforschung* (3 ed.). Oldenbourg Verlag: München

Schorr JE. 1998. *Purchasing in the 21st Century. A Guide to State-of-the-Art Techniques and Strategies*. Wiley: New York, NY

Schreyögg G. 2000. Strategisches Management im Lichte der neuen Steuerungstheorie. In MK Welge, A Al- Laham, P Kajüter (Eds.), *Praxis des strategischen Managements. Konzepte, Erfahrungen, Perspektiven*: 389-405. Gabler: Wiesbaden

Schreyögg G, Steinmann H. 1985. Strategische Kontrolle. *Zeitschrift für betriebswirtschaftliche Forschung* **37**: 391-410

Scott S. 2004. Selecting Suppliers That Suit Your Business Model, *ISM 89th Annual International Supply Management Conference*: Phoenix, AZ

Sharland A, Eltantawy RA, Giunipero LC. 2003. The Impact of Cycle Time on Supplier Selection and Subsequent Performance Outcomes. *Journal of Supply Chain Management: A Global Review of Purchasing and Supply* **39**(3): 4-12

Sharman R. 2002. Enterprise risk management. *The British Journal of Administrative Management* **31**(3): 26-28

Sheth JN, Sharma A. 1997. Supplier Relationships. Emerging Issues and Challenges. *Industrial Marketing Management* **26**: 91-100

Shoemaker PJH. 1990. Strategy, complexity and economic rent. *Management Science* **3**: 1178-1192

Siemens. 2005. Supplier Management (3.1). Siemens AG: München

Silveri RA. 1996. Managing High Technology Suppliers, *NAPM's 81st Annual International Conference*: Chicago, IL

Simon HA (Ed.). 1957. *A Study of Decision-making Processes in Administrative Organizations*. The Free Press: New York, NY

Slack N. 1983. Flexibility as a manufacturing objective. *International Journal of Operations & Production Management* **3**(3): 4-13

Slater SF. 1997. Developing a Customer Value-Based Theory of the Firm. *Journal of the Academy of Marketing Science* **25**(2): 162–167

Slotegraaf RJ, Dickson PR. 2004. The Paradox of a Marketing Planning Capability. *Academy of Marketing Science* **32**(4): 371-385

Smeltzer LR, Siferd SP. 1998. Proactive Supply Management: The Management of Risk. *International Journal of Purchasing and Materials Management* **34**(1): 38-45

Smith KG, Carroll SJ, Ashford SJ. 1991. Intra- and interorganizational cooperation. Toward a research agenda. *Academy of Management Journal* **38**: 7-23

Smith-David J, Hwang Y, Pei BKW, Reneau JH. 1999. *The Impact of Puchasing on Financial Performance*. CAPS: Tempe, AZ

Snow CC, Hrebiniak LG. 1980. Strategy, Distinctive Competence, and Organizational Performance. *Administrative Science Quaterly* **25**: 317-336

Spanos YE, Lioukas S. 2001. An examination into the causal logic of rent generation: Contrasting Porter's competitive strategy framework and the resource-based perspective. *Strategic Management Journal* **22**: 907-923

Speckman RB, Kamauff J. 2002. Supply chain competence: learning as a key component. *Supply Chain Management: An International Journal* **71**: 41-55

Spekman RE. 1985. Competitive Procurement Strategies: Building Strength and Reducing Vulnerability. *Long Range Planning* **18**(1): 94-99

Spekman RE, Davis EW. 2004. Risky business: expanding the discussion on risk and the extended enterprise. *International Journal of Physical Distribution & Logistics Management* **34**(5): 414-433

Spekman RE, Kamauff J, Myhr N. 1998. An empirical investigation into supply chain management: a perspective on partnerships. *Supply Chain Management Review* **3**(2): 53-58

Spekman RE, Kamauff J, Spear J. 1999. Towards more effective sourcing and supplier management. *European Journal of Purchasing & Supply Management* **5**: 103-116

Stalk G, Evans P, Shulman LE. 1992. Competing on Capabilities: The New Rules of Corporate Strategy. *Harvard Business Review* **70**(2): 57-69

Stauffer D. 2003. Risk: The Weak Link in Your Supply Chain, *Harvard Management Update*: 3-5: Boston, MA

Stier W. 1999. *Empirische Forschungsmethoden* (2 ed.). Springer: Berlin

Stölzle W, Gareis K. 2002. Konzepte der Beschaffungslogistik - Anforderungen und Gestaltungsalternativen. In D Hahn, L Kaufmann (Eds.), *Handbuch industrielles Management*, 2 ed.: 402-422. Gabler: Wiesbaden

Stone M. 1975. Cross-validatory Choice and Assessment of Statistical Predictions. *Journal of the Royal Statistical Society* **36**(2): 111-133

Stover LL. 1998. Strategy Integration: Leveraging Opportunities With R&D New Product Teams, *85th NAPM Annual International Purchasing Conference*: Phoenix, AZ

Stuart IF. 1997. Supply-Chain Strategy: Organizational Influence Through Supplier Alliances. *British Journal of Management* **8**: 223-236

Stueland VJ. 2004. Supplier Evaluations: Best Practices and Creating or Improving Your Own Evaluation, *ISM 89th Annual International Supply Management Conference*: Phoenix, AZ

Sun H, Hong CH. 2002. The alignment between manufacturing and business strategies: its influence on business performance. *Technovation* **22**: 699-705

Swift CO. 1995. Preferences for Single Sourcing and Supplier Selection Criteria. *Journal of Business Research* **32**(2): 105-111

Szwejczewski M, Goffin K, Lemke F, Pfeiffer R, Lohmüller B. 2001. Supplier management in German manufacturing companies - an empirical investigation. *International Journal of Physical Distribution & Logistics Management* **31**(5): 354-373

Tallon PP, Kraemer KL. 2002. The Impact of IT Capabilities on Firm Performance: Perspectives on the Mediating Effects of Strategic Alignment. Wallace E. Carroll School of Management. Boston College: Chestnut Hill, MA

Talluri S, Narasimhan R. 2004. A methodology for strategic sourcing. *European Journal of Operational Research* **154**: 236-250

Tamas M. 2000. Mismatched strategies: the weak link in the supply chain? *Supply Chain Management: An International Journal* **5**(4): 171-175

Tan KC, Kannan VR, Handfield RB. 1998. Supply Chain Management: Supplier Performance and Firm Performance. *International Journal of Purchasing and Materials Management* **34**(3): 2-9

Teece DJ, Pisano G, Shuen A. 1997. Dynamic capabilities and strategic management. *Strategic Management Journal* **18**: 509-533

Testore R. 1998. World class manufacturing demands world class suppliers. *European Journal of Purchasing & Supply Management* **4**: 3-5

Thompson AA, Strickland AJ. 1990. *Strategic management: concepts and cases* (5 ed.). McGraw-Hill: Homewood, IL

Thomsen E-H. 2001. *Management von Kernkompetenzen*. Verlag Wissenschaft & Praxis: Sternenfels

Thomson JB, Anderson JC. 2000. Pursuing risk-sharing, gain-sharing arrangements. *Marketing Management* **9**(2): 40-47

Towill D, Christopher M. 2002. The Supply Chain Strategy Conundrum: To be Lean Or Agile or To be Lean AND Agile? *International Journal of Logistics: Research and Applications* **5**(3): 299-309

Trent RJ. 2004. What Everybody Needs to Know about SCM. *Supply Chain Management Review* **8**(3): 52-59

Trent RJ, Monczka RM. 1999. Achieving world-class supplier quality. *Total Quality Management* **10**(6): 927-938

Trent RJ, Monczka RM. 2005. Achieving Excellence in Global Sourcing. *MIT Sloan Management Review* **47**(1): 24-32

Trienekens JH, Beulens AJM. 2001. Views on inter-enterprise relationships. *Production Planning & Control* **12**(5): 466-477

Turner I. 1988. An independent system for the evaluation of contract tenders. *Journal of Operational Research Society* **39**(6): 551–561

Ulaga W. 2003. Capturing value creation in business relationships: A customer perspective. *Industrial Marketing Management* **32**: 677-693

Ulrich H. 1984. *Management*. Paul Haupt: Bern

Ulrich H. 1985. Organisation und Organisieren in der Sicht der systemorientierten Managementlehre. *Zeitschrift für Organisation und Führung* **54**(1): 7-11

Ulrich H. 1995. Von der Betriebswirtschaftslehre zur systemorientierten Managementlehre. In R Wunderer (Ed.), *Betriebswirtschaftslehre als Management- und Führungslehre*, 3 ed.: 161-178. Schäffer-Poeschel: Stuttgart

Ulrich H, Probst G. 1984. *Self-Organization and Management of Social System*. Springer: Berlin

van Hoek RI, Harrison A, Christopher M. 2001. Measuring agile capabilities in the supply chain. *International Journal of Operations & Production Management* **21**(1/2): 126-147

van Weele AJ. 1984. Purchasing Performance Measurement and Evaluation. *Journal of Purchasing and Materials Management* **20**(3): 16-22

van Weele AJ. 1994. *Purchasing Management. Analysis, Planning and Practice*. Chapman & Hall: London

van Weele AJ. 2002. *Purchasing and Supply Chain Management. Analysis, Planning and Practice* (3 ed.). Thompson: London

van Weele AJ, Rozemeijer FA. 1996. Revolution in purchasing. Building competitive power through proactive purchasing. *European Journal of Purchasing & Supply Management* **2**: 153-160

Varadarajan PR. 2003. Musings on Relevance and Rigor of Scholarly Research in Marketing. *Journal of the Academy of Marketing Science* **31**(4): 368-376

Venaik S, Midgley DF, Gable GG. 1999. A Structural Model of Software Maintainer Effectiveness, *10th Australasian Conference on Information Systems*: Wellington

Venkatraman N, Grant JH. 1986a. Construct Measurement in Organizational Strategy Research: A Critique and Proposal. *Academy of Management Review* **11**(1): 71-87

Venkatraman N, Ramanujam V. 1986b. Measurement of business performance in strategy research: a comparison of approaches. *Academy of Management Review* **21**(5): 523-539

Vickery SK, Droge C, Jayaram J, Calantone R. 2003. The effects of an integrative supply chain strategy on customer service and financial performance: an analysis of direct versus indirect relationships. *Journal of Operations Management* **21**: 523-539

Virolainen V-M. 1998. A survey of procurement strategy development in industrial companies. *International Journal of Production Economics* **56-57**: 677-688

Vollmann T, Berry W, Whybark D. 1984. *Manufacturing, Planning and Control Systems*. Dow Jones-Irwin: Homewood; IL

Vonderembse MA. 2002. *Building Supplier Relations That Enhance Manufacturing Performance*. Spiro: London

Vonderembse MA, Tracey M. 1999. The Impact of Supplier Selection Criteria and Supplier Involvement on Manufacturing Performance. *The Journal of Supply Chain Management: A Global Review of Purchasing and Supply* **35**(3): 33-39

Wagner SM. 2001. *Strategisches Lieferantenmanagement in Industrieunternehmen. Eine empirische Untersuchung von Gestaltungskonzepten*. Lang: Frankfurt am Main

Wagner SM. 2003. Intensity and managerial scope of supplier integration. *Journal of Supply Chain Management* **39**(4): 4-15

Wagner SM, Johnson JL. 2004. Configuring and managing strategic supplier portfolios. *Industrial Marketing Management* **33**: 717-730

Walter A, Ritter T, Gemünden HG. 2001. Value Creation in Buyer-Seller Relationships. Theoretical Considerations and Empirical Results from a Supplier's Perspective. *Industrial Marketing Management* **30**: 365-377

Ward PT, Bickford DJ, Leong GK. 1996. Configurations of Manufacturing Strategy, Business Strategy, Environment and Structure. *Journal of Management* **22**(4): 597-626

Watts CA, Kim KY, Hahn CK. 1995. Linking Purchasing to Corporate Competitive Strategy. *Journal of Supply Chain Management* **31**(2): 2-8

Weigand M. 1997. *Die Erschließung von Zulieferpotentialen als Aufgabe des strategischen Beschaffungsmarketing*: Erlangen-Nürnberg

Weissman RG. 2004. Creating Customer Value Through Supplier Performance, *ISM 89th Annual International Supply Management Conference*: Phoenix, AZ

Welge MK, Al-Laham A. 1992. *Planung, Prozesse – Strategien – Maßnahmen*. Gabler: Wiesbaden

Wen-li L, Humphreys P, Chan L, Kumaraswamy M. 2003. Predicting purchasing performance: the role of supplier development programs. *Journal of Materials Processing Technology* **138**: 243-249

Wernerfelt B. 1984. A resource-based view of the firm. *Strategic Management Journal* **5**: 171-180

Westermann H. 1998. SWOT-Analyse im Einkauf, *Beschaffung Aktuell*: 54-58

Whetten DA. 1989. What Constitutes a Theoretical Contribution? *Academy of Management Review* **14**(4): 490-495

Whipple JM, Frankel R. 2000. Strategic Alliance Success Factors. *The Journal of Supply Chain Management: A Global Review of Purchasing and Supply* **36**(3): 21-28

William G, Katircioglu K, Shi D, Bagchi S, Gallego G, Adelhelm M, Seybold D, Stefanis S. 2003. Beyond ROI. *Supply Chain Management Review* **7**(2): 20-27

Wilson DT. 1995. An Integrated Model of Buyer-Seller Relationships. *Journal of the Academy of Marketing Science* **23**: 335-345

Wisner JD, Tan KC. 2000. Supply Chain Management and its Impact on Purchasing. *Journal of Supply Chain Management* **36**(4): 33-42

Wohlgemuth O. 2002. *Management netzwerkartiger Kooperationen. Instrumente für die unternehmensübergreifende Steuerung*. DUV: Wiesbaden/Dissertation University of Göttingen

Wold H. 1975. Path models with latent variables: The NIPALS approach. In HM Blalock, A Aganbegian, FM Borodkin, R Boudon, V Capecchi (Eds.), *Quantitative sociology: International perspectives on mathematical and statistical modeling*: 307-357. Academic: New York, NY

Wold H. 1982. Soft modeling: the basic design and some extensions. In KG Jöreskog, H Wold (Eds.), *Systems under indirect observations: Causality, structure, precision*: 1-54. North-Holland: Amsterdam

Wolters P. 2002. Foward Sourcing - Entwicklungsbegleitende Lieferantenauswahl. In D Hahn, L Kaufmann (Eds.), *Handbuch industrielles Management*, 2 ed.: 340-363. Gabler: Wiesbaden

Womack JP, Jones DT, Roos D. 1990. *The machine that changed the world*. Harper Perennial: New York

Woodruf RB, Gardial SF. 1996. *Know Your Customer: New Approaches to Customer Value and Satisfaction*. Cambridge University Press: Cambridge

Young A, Kielkiewicz-Young A. 2001. Sustainable Supply Network Management. *Corporate Environmental Strategy* **8**(3): 260-268

Yuva J. 2004. Cross-Examing Business Trends, *Inside Supply Management*, Vol. 15: 22-30

Yuva J, Duffy RJ. 2002. The Revenue Stream Flows from Supply Management, *Inside Supply Management*, Vol. 13: 34-38

Zahra SA, Covin JG. 1993. Business Strategy, Technology Policy and Firm Performance. *Strategic Management Journal* **14**: 451-478

Zajac EJ, Olsen CP. 1993. From transaction cost to transactional value analysis: Implications for the study of interorganizational strategies. *Journal of Management Studies* **30**(1): 131-145

Zinnbauer M, Eberl M. 2004. Die Überprüfung von Spezifikation und Güte von Strukturgleichungsmodellen: Verfahren und Anwendung., *Schriften zur Empirischen Forschung und Quantiativen Unternehmensplanung. Heft 21*. LMU München: München

Zollenkop M. 2003. Global Sourcing. In H-J Bullinger, HJ Warnecke, E Westkämper (Eds.), *Neue Organisationsformen in Unternehmen*, 2 ed.: 587-597. Springer: Berlin

Zsidisin GA. 2003. Managerial Perceptions of Supply Risk. *The Journal of Supply Chain Management: A Global Review of Purchasing and Supply* **39**(1): 14-25

Zsidisin GA, Ellram LM, Odgen JA. 2003. The relationship between purchasing and supply management's perceived value and participation in strategic supplier cost management activities. *Journal of Business Logistics* **24**(2): 129-154

Zsidisin GA, Panelli A, Upton R. 2000. Purchasing organization involvement in risk assessments, contingency plans, and risk management: an explorative study. *Supply Chain Management* **5**(4): 187-196